Differentiable Dynamical Systems

An Introduction to Structural Stability and Hyperbolicity

GRADUATE STUDIES IN MATHEMATICS 173

Differentiable Dynamical Systems

An Introduction to Structural Stability and Hyperbolicity

Lan Wen

American Mathematical Society
Providence, Rhode Island

2010 *Mathematics Subject Classification.* Primary 37Cxx, 37Dxx, 37Bxx, 34Cxx, 34Dxx.

For additional information and updates on this book, visit
www.ams.org/bookpages/gsm-173

Library of Congress Cataloging-in-Publication Data

Names: Wen, Lan, 1946–
Title: Differentiable dynamical systems : an introduction to structural stability and hyperbolicity / Lan Wen.
Description: Providence, Rhode Island : American Mathematical Society, [2016] | Series: Graduate studies in mathematics ; volume 173 | Includes bibliographical references and index.
Identifiers: LCCN 2016012111 | ISBN 9781470427993 (alk. paper)
Subjects: LCSH: Differential equations. | Differential equations, Nonlinear. | Dynamics. | Exponential functions. | AMS: Dynamical systems and ergodic theory – Smooth dynamical systems: general theory – Smooth dynamical systems: general theory. msc | Dynamical systems and ergodic theory – Dynamical systems with hyperbolic behavior – Dynamical systems with hyperbolic behavior. msc | Dynamical systems and ergodic theory – Topological dynamics – Topological dynamics. msc | Ordinary differential equations – Qualitative theory – Qualitative theory. msc | Ordinary differential equations – Stability theory – Stability theory. msc
Classification: LCC QA371 .W495 2016 | DDC 515/.39–dc23 LC record available at http://lccn.loc.gov/2016012111

♾ The paper used in this book is acid-free and falls within the guidelines
established to ensure permanence and durability.
Visit the AMS home page at `http://www.ams.org/`

10 9 8 7 6 5 4 3 2 1 21 20 19 18 17 16

To my colleagues and students

Contents

Preface

Dynamical systems began with Poincaré's qualitative theory of differential equations near the end of the nineteenth century. Differentiable dynamical systems is the part that concerns structural stability, hyperbolicity, genericity, density, etc., developed since the 1960s. There is a survey article by Smale (1967, also 1980) on differentiable dynamical systems that is very enlightening.

This book is a graduate text in differentiable dynamical systems. It focuses on structural stability and the role of hyperbolicity, a topic that is central to the field. For the sake of simplicity we take the discrete setting, namely iterates of diffeomorphisms. It is well known that a periodic orbit is structurally stable if and only if it obeys so-called hyperbolicity (no eigenvalue of norm one). The case of a finite number of periodic orbits is treated similarly. However, it was long doubted whether a system of infinitely many periodic orbits could be structurally stable. A historic discovery from the early 1960s is the Smale horseshoe map. This is a structurally stable system that contains infinitely many periodic orbits. Together with the celebrated Anosov automorphism and the solenoid attractor found soon afterwards, they exhibit an amazing feature of the world: structural stability can be compatible to a high level of complexity (sometimes called "chaos"). The analytic condition that ensures such a chaotic set to be structurally stable is reasonably still called "hyperbolicity". This led to a new theory of *hyperbolic sets*, for which a hyperbolic periodic orbit serves as a special case. The Ω-stability theorem of Smale is then an early global result based on this theory. This book will develop along this line and consists of six chapters.

Chapter 1 introduces some basic concepts of dynamical systems such as limit set, nonwandering set, minimal set, transitive set, etc., as well as topological conjugacy and structural stability. As a comprehensive illustration of these concepts we give a short account of the classical theory of circle homeomorphisms. We also include in this chapter Conley's fundamental theorem of dynamical systems.

Chapter 2 is devoted to hyperbolicity, the main analytic concept of this book, for the case of a single fixed point. We study the stability of a hyperbolic fixed point, the persistence of hyperbolicity under perturbations, the Hartman-Grobman theorem, the stable manifold theorem, etc. While these subjects are classical, our treatment has kept in mind the need of the general case of a hyperbolic set of Chapter 4.

Chapter 3 presents three historic models, the Smale horseshoe, the Anosov toral automorphism, and the solenoid attractor, which led to the modern theory of differentiable dynamical systems.

Chapter 4 generalizes the concept of hyperbolicity from a fixed point to a general invariant set. We study the persistence of hyperbolicity, the stable manifold theorem, structural stability, the shadowing property, etc., for a hyperbolic set. This chapter constitutes the analytic foundation for the theory of structural stability and is technically the most difficult part of the book. Whenever some difficulty appears, the best way is to go back to check the corresponding part of Chapter 2, which is much more transparent.

Chapter 5 presents one direction of the theory: hyperbolicity implies (essentially) structural stability. A highlight is the Ω-stability theorem of Smale. We also include some equivalent descriptions by Newhouse and Franke-Selgrade.

Chapter 6 presents the theory of quasi-hyperbolicity and linear transversality. It provides alternate angles for looking at hyperbolicity. We also include a section that gives a glimpse of the stability conjectures.

There is some tough material in this book, notably the stable manifold theorem (Theorem 4.16) and the structural stability theorem (Theorem 4.21) for hyperbolic sets. In fact these two big theorems have caused an obstacle to teaching and learning the subject. A key objective of this book is to find some way to remove this obstacle. The strategy we found is to choose a suitable setting for proofs for a hyperbolic set so that they match the proofs for a hyperbolic fixed point, like a copy. A good example showing that this is possible is the proof of Lemma 4.5, which is almost a duplicate of the proof of Lemma 2.9. The reader might like to take one minute to compare the two proofs just formally. Indeed, with this strategy we have been able to give straightforward proofs for these two big theorems. The author believes that, unlike in art or in literature, in mathematics one does not have to avoid analogous presentations; rather, one uncovers the identical nature behind different exteriors. After all, the theory of hyperbolic sets is difficult to digest. We wish to make the presentation as plain as possible so the reader can quickly get through the obstacle to reach the heart of differentiable dynamical systems.

The prerequisites for reading this book are essentially undergraduate analysis, linear algebra, and basic topology. The framework of differentiable manifolds, in particular some basic concepts such as tangent bundles and tangent maps, submanifolds, Riemannian metric, the exponential map, are also important to the development of the text. When some less standard facts of basic topology are needed, definitions are inserted as a refresher. The book includes a number of figures to help the exposition.

There have been a number of books on dynamical systems in the literature. See the references at the end of this book, where the two big books Katok-Hasselblatt (1995) and Robinson (1995) give a panorama for modern dynamical systems. I have benefited from these nice books in many ways. I am especially thankful for the book by Zhang (1986), which I used as a text in the early years when I taught the course at Peking University.

This book is short. It might miss many original references for the results involved in the text. I am thankful for the book by Robinson (1995), which I found to be very helpful.

I have taught most parts of the first five chapters of the book as a one-semester course many times at Peking University. I also taught the course at Taiwan University (spring 2003) and Providence University of Taiwan (fall 2004). Part of the material was taught as short courses at Nankai University (1989), Sun Yat-Sen University (1990), Fuzhou University (1995), Nanjing University (1998), Center of Theoretical Sciences of Taiwan (1999), University of Science and Technology of China (2001), Jilin University (2007), Chiao Tung University of Taiwan (2011), and Chungnam University of Korea (2014). I wish to take this opportunity to thank the participants of all these courses. I particularly thank Shaobo Gan for collaborations over the years, countless discussions about the course, including a thorough proof-reading of the entire manuscript. I also thank Xiao Wen and Dawei Yang for creating elegant figures and many of the exercises for this book and Xiao Wen for discussions on the C^k part of the stable manifold theorem. Finally, I wish to thank Wenxiang Sun and the people in our seminar for stimulating talks and discussions over the years.

Lan Wen
Peking University

Chapter 1

Basics of dynamical systems

Informally, a dynamical system is just an abstract flow, that is, an abstract motion of points with time. The motion of a single point gives an orbit. The classical example of a dynamical system is a flow determined by an ordinary differential equation (or vector field). As part of the definition of a flow, every solution is defined on the whole line $(-\infty, \infty)$; hence points would never move out. The most special orbit consists of a single point called a singularity of the vector field, for instance a sink, or a source, or a saddle. If some nearby points approach it exponentially with the flow, the singularity will be called hyperbolic. Another sort of special orbits is the periodic ones. If nearby points approach it exponentially, the periodic orbit will be called hyperbolic.

Every scientific discipline has many interesting stories. In 1962, upon a suggestion by Lefschetz, Peixoto revived the pioneer work of Andronov-Pontryagin (1937) on structurally stable vector fields of the 2-disc and proved a new theorem: A vector field on an orientable closed surface is structurally stable if and only if it satisfies three conditions: (1) it has at most finitely many singularities and periodic orbits, each hyperbolic; (2) every point approaches, positively as well as negatively, a single singularity or a single periodic orbit; (3) there is no saddle connection. Moreover, structurally stable vector fields are dense in the space of all vector fields.

Here a vector field is called *structurally stable* if all nearby vector fields have topologically equivalent orbital structures. From any point of view, structural stability is a concept of paramount importance. It is however abstract and involves nearby vector fields, which is difficult to handle. In contrast, the criterion stated in Peixoto's theorem involves merely one vector field and focuses on singularities and periodic orbits only. This beautiful global theorem immediately attracted people's attention, a young mathematician S. Smale in particular. Some years later he wrote down the following interesting story in an article "On how I got started in Dynamical Systems", collected in his booklet *The Mathematics of Time* (1980):

Reading Peixoto's paper, at first Smale thought that the result also holds in higher dimensions. But then Levinson wrote him telling him that one could not expect the result to hold so generally. Levinson (1949) had a 3-dimensional counterexample already in which there are infinitely many periodic orbits that cannot be perturbed away. Still partly with disbelief, Smale spent a lot of time (on a beach of Rio with a pen and a pad of paper) studying Levinson's paper, eventually becoming convinced. In fact Smale abstracted the following geometrical mechanism of what Levinson and Cartwright-Littlewood (1945) had found more analytically before:

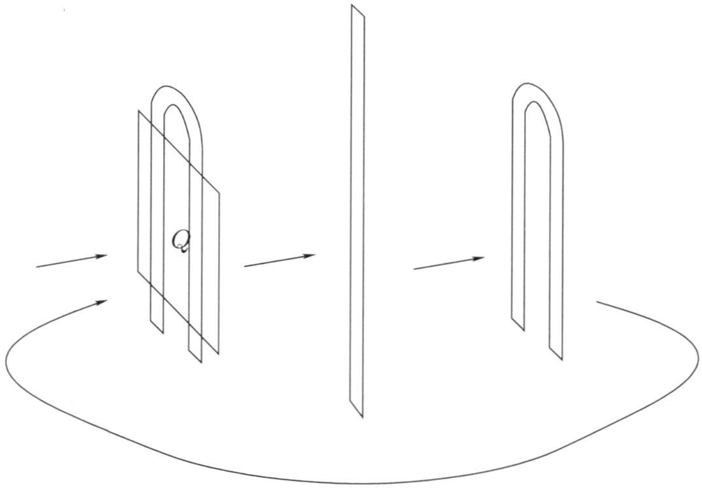

Figure 1.1. A key mechanism in the 3D flow.

As shown in Figure 1.1, Q is a square transverse to the flow direction. Under the flow, Q gets longer and thinner, bending like a horseshoe, and turns back to cross Q itself. (Here we have simplified the story a little bit. For more details see Smale (1980).) Smale realized that it is this simple mechanism that causes the persistent appearance of infinitely many periodic orbits. To see this more clearly, he reduced the 3D flow problem to a 2D map problem. That is, he counts "one turn" of the flow to be a map $f : Q \to \mathbb{R}^2$, as Figure 1.2 shows:

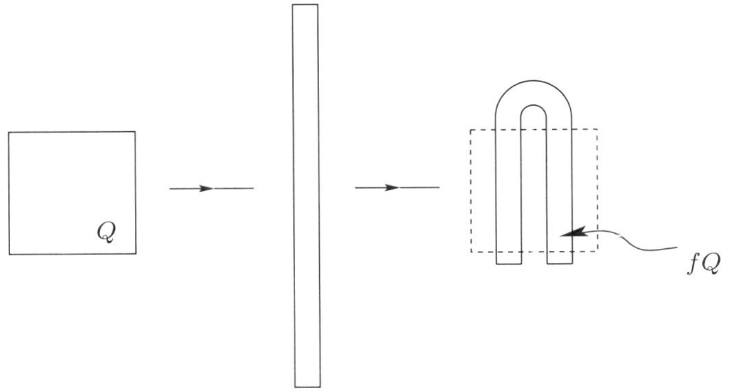

Figure 1.2. The 2D horseshoe map of Smale.

Thus a fixed point of f corresponds to a periodic orbit of the flow. A periodic point of f of period 2 also corresponds to a periodic orbit of the flow, going around twice. Smale proved that f has infinitely many periodic points that cannot be perturbed away. This means that Peixoto's theorem does not hold in higher dimensions. In higher dimensions, structural stability can coexist with a high level of complexity (sometimes called "chaos"). This phenomenon becomes a symbol of the modern differentiable dynamical systems. We will study this horseshoe system in Chapter 3.

Smale soon realized that this problem has a long history. It goes back to the striking homoclinic phenomenon discovered by Poincaré in his study of celestial mechanics and also relates to the work of Birkhoff on surface diffeomorphisms. Based on the horseshoe map and the important work of Anosov (1967) that appeared soon afterwards, Smale came up with the notion of hyperbolic set that contains a hyperbolic periodic orbit as a special case and formulated with Palis (1970) the higher-dimensional version of Peixoto's criteria, the *stability conjectures*, which led to many important works, pushing differentiable dynamical systems to a full flourish.

As to the density problem, the situation is quite different. Starting with Smale (1966), one after another, many authors found in a more and more striking sense that, in higher dimensions, structurally stable systems are simply not dense. That is, there is an open set like a "strange hole" in the space of all systems such that every system in the hole is not structurally stable. Bonatti and Diaz (2003) even found a hole $\mathcal{U}$ in the space of all systems with *universal dynamics* in the sense that, contrary to the situation of unique dynamics near any structurally stable system, here, arbitrarily near every system in $\mathcal{U}$, up to suitable iterates, any system can occur! Dynamical systems seems to be destined to surprise us. To grasp the general feature in the holes, Palis (2000, 2005) proposed since the 1990s a number of stimulating conjectures, notably the *density conjectures*, stating that some specific unstable dynamics called homoclinic bifurcations are typical in all possible holes, which have led to many exciting works. The mystery of nature, the wonder of nature, are uncovered this way, step by step, by explorers of different eras ....

This book presents a first but fundamental part of this beautiful field. For simplicity we consider systems of discrete time only, that is, iterates of diffeomorphisms.

1.1. Basic concepts

Let X be a compact metric space and $f : X \to X$ be a homeomorphism. This generates a family of homeomorphisms, called *iterates* of f, written as

$$f^n = \underbrace{f \circ f \circ \cdots \circ f}_{n}, \quad f^0 = id, \quad f^{-n} = (f^n)^{-1}.$$

It is obvious that

$$f^n \circ f^m = f^{n+m}$$

for any integers n and m. We call the family $\{f^n\}_{n=-\infty}^{\infty}$ a *dynamical system*, or we simply call f a dynamical system.

For any $x \in X$, the set $\{f^n(x)\}_{n=-\infty}^{\infty}$ is called the *orbit* of x under f, denoted by $\text{Orb}(x, f)$, or simply by $\text{Orb}(x)$. See Figure 1.3. Any two orbits are either identical or else disjoint. The sets $\{x, fx, f^2x, \ldots\}$ and $\{x, f^{-1}x, f^{-2}x, \ldots\}$ are called the *positive orbit* and the *negative orbit* of x and are denoted by $\text{Orb}^+(x)$ and $\text{Orb}^-(x)$, respectively. A point $x \in X$ is called *periodic* if there is $n \geq 1$ such that $f^n(x) = x$. The minimal positive integer n that satisfies this equality is called the *period* of x. The orbit of a periodic point is called a *periodic orbit*. Periodic points of period 1 are just fixed points. It is easy to see that $x \in X$ is periodic if and only if $\text{Orb}(x)$ consists of finitely many points. We denote the set of periodic points of f by $\text{P}(f)$, and we denote the set of fixed points of f by $\text{Fix}(f)$.

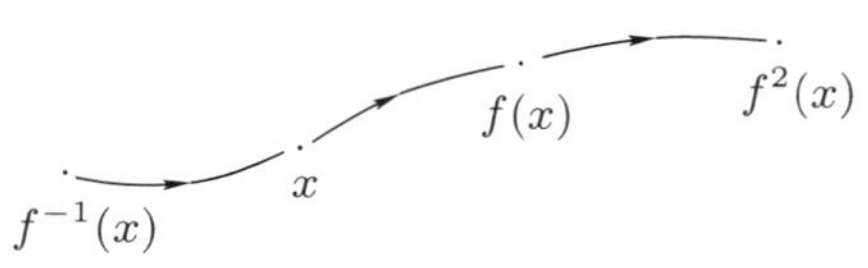

Figure 1.3. Orbit.

A subset $\Lambda \subset X$ is called *invariant* under f if $f(\Lambda) = \Lambda$. Any orbit is invariant. Λ is invariant if and only if Λ is a union of orbits.

Theorem 1.1. *If Λ is invariant, so are $\overline{\Lambda}$, $\partial(\Lambda)$, and* $\mathrm{int}(\Lambda)$.

Proof. Since f is a homeomorphism, we have $f(\overline{\Lambda}) = \overline{f(\Lambda)} = \overline{\Lambda}$. The other two are proved similarly. □

The set $\mathrm{Fix}(f)$ of fixed points is compact and invariant but might be empty. The set $\mathrm{P}(f)$ of periodic points is invariant but might be empty, or nonempty but not compact. Now we give more invariant sets.

Given $x \in X$, the positive orbit $x, fx, f^2x, \dots$ generally does not converge (if it does, the limit must be a fixed point). Nevertheless many subsequences of it do. A point $y \in X$ is called an *ω-limit point* of $x \in X$ if there is a subsequence $n_i \to +\infty$ of the positive integers such that $f^{n_i}(x) \to y$. The set of ω-limit points of x is called the *ω-limit set* of x, denoted by $\omega(x)$. See Figure 1.4. Reversing time defines the α-limit set of x. That is, a point $y \in X$ is called an *α-limit point* of $x \in X$ if there is a subsequence $n_i \to +\infty$ of the positive integers such that $f^{-n_i}(x) \to y$. The set of α-limit points of x is called the *α-limit set* of x, denoted by $\alpha(x)$. Clearly $\alpha(x) = \omega(x, f^{-1})$. One usually states results for $\omega(x)$ only. Note that if $x \in \mathrm{P}(f)$, then

$$\omega(x) = \alpha(x) = \mathrm{Orb}(x).$$

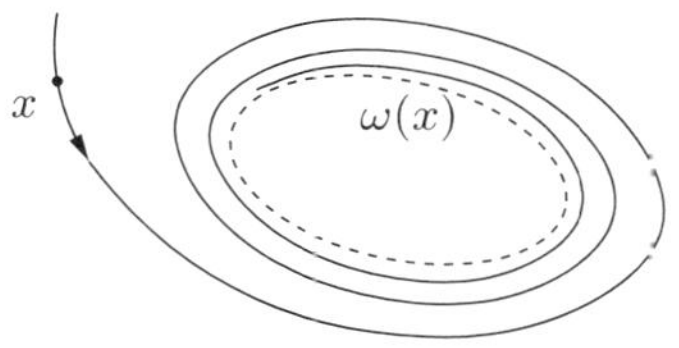

Figure 1.4. ω-limit set.

Theorem 1.2. *For any $x \in X$, $\omega(x)$ is nonempty, compact, and invariant. Moreover,*

$$\lim_{n\to\infty} d(f^n(x), \omega(x)) = 0.$$

Proof. Since X is compact, it follows that $\omega(x)$ is nonempty and compact. Take $y \in \omega(x)$. There is a subsequence $n_i \to +\infty$ such that $f^{n_i}(x) \to y$. Then $f^{n_i+1}(x) \to f(y)$; that is, $f(y) \in \omega(x)$. Thus $f(\omega(x)) \subset \omega(x)$. Likewise, $f^{n_i-1}(x) \to f^{-1}(y)$, hence $f^{-1}(\omega(x)) \subset \omega(x)$. Thus $f(\omega(x)) \supset \omega(x)$. This proves $\omega(x)$ is invariant.

Now suppose the equality $\lim_{n\to\infty} d(f^n(x), \omega(x)) = 0$ does not hold. Then there are $\epsilon_0 > 0$ and a subsequence $n_i \to +\infty$ such that $\mathrm{d}(f^{n_i}(x), \omega(x)) \geq \epsilon_0$ for all i. Taking subsequence n_{i_k} yields $f^{n_{i_k}} \to z \notin \omega(x)$, a contradiction. □

Dynamical systems emphasize the long run behavior of orbits, which is contained in the set

$$\mathrm{L}(f) = \overline{\bigcup_{x \in X} \omega(x) \cup \alpha(x)},$$

called the *limit set* of f. It is nonempty, compact, and invariant.

Dynamical systems also emphasize recurrence of orbits. A periodic orbit has the strongest recurrence. Besides the periodicity, there are several notions of recurrence, each one weaker than the previous one. A point $x \in X$ is called *positively recurrent* if $x \in \omega(x)$. In other words, $x \in X$ is positively recurrent if its positive orbit accumulates on x itself. Likewise, a point $x \in X$ is called *negatively recurrent* if $x \in \alpha(x)$. Positively or negatively recurrent points are both called *recurrent*. Denote by $R(f)$ the set of recurrent points of f. It is nonempty, invariant, but perhaps noncompact (Exercise 1.6).

A still weaker recurrence is the so-called nonwandering property. A point $x \in X$ is called *nonwandering* under f if, for any neighborhood V of x in X, there is $n \geq 1$ such that $f^n(V) \cap V \neq \emptyset$. In other words, for any neighborhood V of x in X, there is some orbit that hits V at least twice. See Figure 1.5. The set of nonwandering points of f is called the *nonwandering set* of f, denoted by $\Omega(f)$. Clearly, $\Omega(f)$ is nonempty, compact, and invariant.

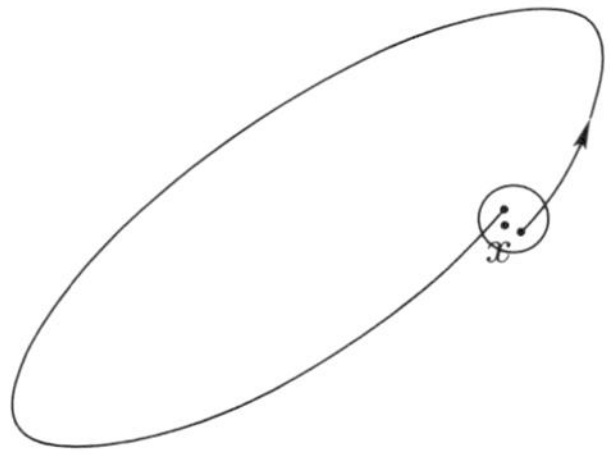

Figure 1.5. Nonwandering point x.

Finally, by an *ϵ-chain from x to y* we mean a finite sequence $x_0, x_1, \ldots, x_k$ with $x_0 = x,\ x_k = y$ such that

$$d(f(x_n),\ x_{n+1}) < \epsilon$$

for $n = 0, 1, \ldots, k-1$. In case $x = y$, the ϵ-chain is called *periodic.* A point $x \in X$ is said to be *chain recurrent* of f if, for any $\epsilon > 0$, there is an ϵ-chain from x to x, or equivalently, if there is a periodic ϵ-chain going through x. The set of chain recurrent points of f is called the *chain recurrent set* of f, denoted $\mathrm{CR}(f)$. In a reasonable sense, chain recurrence is the weakest notion of recurrence. See the remark after Lemma 1.15.

The notion of chain recurrence is flexible. See the exercises at the end of this chapter. For instance, a point $x \in X$ is chain recurrent if and only if for any $\epsilon > 0$, there is a periodic ϵ-chain going through the ϵ-neighborhood of x (Exercise 1.9). It is easy to see $\mathrm{CR}(f)$ is compact invariant and

$$\overline{\mathrm{P}(f)} \subset \mathrm{L}(f) \subset \Omega(f) \subset \mathrm{CR}(f).$$

A nonempty compact invariant set is like a subdynamical system. A set $\Lambda \subset X$ is called *minimal* if Λ is nonempty compact invariant, but no proper subset of Λ is nonempty compact invariant. A periodic orbit is clearly a minimal set. While for some dynamical systems periodic orbits may not exist, the following theorem says minimal sets always exist. The proof uses Zorn's lemma, which we now review.

Let S be a set. Let $\prec$ be a binary relation defined for some of the pairs of elements of S. We say $\prec$ is a *partial order* if (1) $x \prec x$ for every $x \in S$; (2) if $x \prec y$ and $y \prec x$, then $x = y$; (3) if $x \prec y$ and $y \prec z$, then $x \prec z$. We emphasize that some pair (x, y) of elements of S may not be comparable meaning neither $x \prec y$ nor $y \prec x$. A typical partial order is the inclusion of sets, $\subset$, with which we are familiar. A subset A of S is called *totally ordered* (with respect to $\prec$) if for every pair (x, y) of elements of A, either $x \prec y$ or $y \prec x$.

Let $z \in S$. We say z is a *minimal element* of S if, for every $x \in S$, either z and x are not comparable or $z \prec x$. Let $z \in S$ and $A \subset S$. We say z is a *lower bound* of A if $z \prec x$ for every $x \in A$. Zorn's lemma asserts that *if every totally ordered subset A of S has a lower bound, then S has a minimal element.*

Theorem 1.3. *Any nonempty compact invariant set contains a minimal set.*

Proof. Let Γ be a nonempty compact invariant set of f. Let $\mathcal{C}$ be the set of nonempty compact invariant subsets of f contained in Γ. The inclusion $\subset$ is a partial order on $\mathcal{C}$. Let $\mathcal{A}$ be a totally ordered subset of $\mathcal{C}$. Let A be the intersection of all elements of $\mathcal{A}$. Then A is a nonempty compact invariant set of f. Clearly A is a lower bound of $\mathcal{A}$. By Zorn's lemma, $\mathcal{C}$ has a minimal element, which is just a minimal set of f. □

The next lemma states that minimality yields a strong recurrence.

Theorem 1.4. *A compact invariant set Λ is minimal if and only if the orbit of every $x \in \Lambda$ is dense in Λ.*

Proof. Let Λ be minimal. Take any $x \in \Lambda$. Since $\overline{\mathrm{Orb}(x)} \subset \Lambda$ is nonempty compact invariant, by minimality $\overline{\mathrm{Orb}(x)} = \Lambda$. Conversely, assume Λ is not minimal, then there is a proper subset $\Lambda_1 \subset \Lambda$ that is nonempty compact invariant. Take any $x \in \Lambda_1$. Then $\overline{\mathrm{Orb}(x)} \subset \Lambda_1 \neq \Lambda$. □

Recall that a subset Λ of X is called *nowhere dense* in X if the closure $\overline{\Lambda}$ has no interior in X. For instance, a closed interval is nowhere dense in the plane.

Theorem 1.5. *Assume X is connected. Then any minimal set of f is either the whole X or nowhere dense in X.*

Proof. Let Λ be a minimal set of f. Note that $\partial\Lambda$ is compact invariant. If $\partial\Lambda = \emptyset$, then $\Lambda = \text{int}(\Lambda)$. Hence Λ is open. Thus Λ is both open and closed, hence equal to X since X is connected. If $\partial\Lambda \neq \emptyset$, then $\partial\Lambda$ is nonempty compact invariant. By minimality $\partial\Lambda = \Lambda$. Then Λ has no interior in X, meaning Λ is nowhere dense in X. □

We call a compact invariant set Λ *indecomposable* if Λ cannot be decomposed into a disjoint union of two (nonempty) compact invariant sets. Of course a minimal set is indecomposable. A more general property that guarantees the indecomposability is topological transitivity. A compact invariant set $\Lambda \subset X$ of f is called *topologically transitive*, or *transitive*, if there is $x \in \Lambda$ such that $\omega(x) = \Lambda$. Clearly, a transitive set is indecomposable.

Theorem 1.6 (Birkhoff). *Let Λ be a compact invariant set of f. The following conditions are equivalent:*

(1) *Λ is transitive.*

(2) *For any two open subsets U and V of Λ, there is $n \geq 1$ such that $f^n(U) \cap V \neq \emptyset$.*

(3) *There is $x \in \Lambda$ whose positive orbit is dense in Λ.*

Proof. The proof of (1)⇒(2) is easy, hence omitted. We prove (2) ⇒ (3). Take a countable basis $V_1, V_2, \ldots$ of Λ. For any $i \geq 1$, the set $\bigcup_{n=1}^{\infty} f^{-n}V_i$ is open in Λ. It is dense in Λ as well because, for any open set U in Λ, there is by (2) $n \geq 1$ such that $f^n(U) \cap V_i \neq \emptyset$; that is, $U \cap f^{-n}(V_i) \neq \emptyset$. By Baire's theorem,

$$B = \bigcap_{i=1}^{\infty} \bigcup_{n=1}^{\infty} f^{-n}V_i$$

is nonempty (in fact dense in Λ). Take any $x \in B$. Then for any $i \geq 1$, there is $n \geq 1$ such that $x \in f^{-n}V_i$; that is, $f^n x \in V_i$. This proves that $\text{Orb}^+(x)$ is dense in Λ, proving (2) ⇒ (3).

Now we prove (3) ⇒ (1). Assume there is $x \in \Lambda$ such that $\Lambda = \overline{\text{Orb}^+(x)}$. Then $f^{-1}(x) \in \overline{\text{Orb}^+(x)}$. If $f^{-1}(x) \in \text{Orb}^+(x)$, then x is periodic. Hence $\Lambda = \omega(x)$. If $f^{-1}(x) \notin \text{Orb}^+(x)$, then $f^{-1}(x) \in \omega(x)$. Hence $\text{Orb}(x) \subset \omega(x)$. Then $\Lambda = \overline{\text{Orb}^+(x)} \subset \omega(x)$. But Λ contains x and hence contains $\omega(x)$. Hence $\Lambda = \omega(x)$. This proves Theorem 1.6. □

Remark. In condition (3), "positive" could be replaced by "negative". Sometimes condition (3) is incorrectly stated as "there is $x \in \Lambda$ whose orbit is dense in Λ", or simply as "there is a dense orbit", without emphasizing the word "positive" (or "negative"). A counterexample Λ consists of two fixed points plus an orbit O that "connects" them. Then Λ has a dense orbit O but is not transitive. Thus the word positive (or negative) is necessary.

A natural intention is to decompose a compact invariant set into a disjoint union of (even if infinitely many) indecomposable compact invariant sets. The following equivalence relation provides such a decomposition for the chain recurrent set $\mathrm{CR}(f)$.

Two points $x, y \in \mathrm{CR}(f)$ are called *chain equivalent*, written $x \sim y$, if for any $\epsilon > 0$, there are an ϵ-chain from x to y and an ϵ-chain from y to x. This is an equivalence relation on $\mathrm{CR}(f)$. Each equivalence class is called a *chain transitive class*, or simply *chain class* of f. Each chain class is compact and f-invariant.

The next theorem says that, when $\epsilon \to 0$, periodic ϵ-chains going through a point of a chain class are forced to gather round the chain class, and hence any chain class is indecomposable.

Theorem 1.7. *Let C be a chain class of f. Then*

(1) *for any $\epsilon > 0$ there is $\delta > 0$ such that, for any $x \in C$, any periodic δ-chain through x is contained in the ϵ-neighborhood $B(C, \epsilon)$ of C;*

(2) *C is indecomposable.*

Proof. (1) Suppose there is $\epsilon_0 > 0$ such that, for every $n \geq 1$, there is $x_0^n \in C$ and a periodic $1/n$-chain

$$x_0^n,\ x_1^n,\ \ldots,\ x_{j_n}^n$$

such that $x_{k_n}^n \notin B(C, \epsilon_0)$ for some k_n. Taking subsequences if necessary, we may assume $x_0^n \to x$, $x_{k_n}^n \to y$. Then x is chain equivalent to y. But $x \in C$ and $y \notin C$, a contradiction.

(2) Suppose C decomposes into a disjoint union of two (nonempty) compact invariant sets C_1 and C_2. Take $\epsilon > 0$ small such that

$$B(C_1, \epsilon) \cap B(C_2, \epsilon) = \emptyset, \quad (fB(C_1, \epsilon)) \cap B(B(C_2, \epsilon), \epsilon) = \emptyset.$$

The second equality means that any point in $B(C_1, \epsilon)$ cannot jump in one step into the ϵ-neighborhood of $B(C_2, \epsilon)$.

Take $x \in C_1$. By (1), there is $\delta > 0$ such that any periodic δ-chain through x is contained in $B(C_1, \epsilon) \cup B(C_2, \epsilon)$. We may assume $\delta < \epsilon$. Since C_1 and C_2 are in the same chain class, there is a periodic δ-chain through

x that hits $B(C_2, \epsilon)$. Then there is a point $z \in B(C_1, \epsilon)$ such that $f(z) \in B(B(C_2, \epsilon), \delta) \subset B(B(C_2, \epsilon), \epsilon)$, contradicting the choice of ϵ. □

Thus, $\mathrm{CR}(f)$ decomposes into a disjoint union of indecomposable pieces, the chain classes. A system may have infinitely many chain classes. The simplest example is a homeomorphism of $[0, 1]$ with infinitely many fixed points p_n, which are alternate sinks and sources, accumulating to the fixed point 0. Each p_n is a chain class.

1.2. Topological conjugacy and structural stability

Two homeomorphisms $f : X \to X$ and $g : X \to X$ are called *topologically conjugate* to each other if there is a homeomorphism $h : X \to X$ such that $hf = gh$. Roughly, two such f and g differ by a continuous change of coordinates. The homeomorphism h is called a *topological conjugacy*, or simply a *conjugacy* from f to g. Being conjugate to each other is an equivalence relation on the space of all homeomorphisms. It is easy to see that $hf^n = g^n h$. Hence a conjugacy h preserves orbits; that is,

$$h(\mathrm{Orb}(x, f)) = \mathrm{Orb}(h(x), g)$$

for any $x \in X$. In particular, a conjugacy preserves the periodic set, ω-limit sets, nonwandering set, and the chain recurrent set. That is,

$$h(\mathrm{P}(f)) = \mathrm{P}(g), \ h(\omega(x, f)) = \omega(h(x), g),$$

$$h(\Omega(f)) = \Omega(g), \ h(\mathrm{CR}(f)) = \mathrm{CR}(g).$$

To classify all homeomorphisms up to conjugacy would be ideal, but not realistic. Nevertheless for extremely simple spaces such a classification is not hard. Let us take the case that X is a closed interval $[a, b]$. A homeomorphism $f : [a, b] \to [a, b]$ is either strictly increasing and fixes the two end points of the interval or else strictly decreasing and interchanges the two end points of the interval. In the first case f is called *orientation preserving*. In the second case f is called *orientation reversing*. An orientation-preserving homeomorphism cannot be conjugate to an orientation-reversing one because otherwise they would be conjugate on the boundary $\partial([a, b])$. However, restricted to $\partial([a, b])$, an orientation-preserving homeomorphism has two fixed points, while an orientation-reversing one has none.

Interval homeomorphisms have the simplest recurrence. Precisely, for any orientation-preserving homeomorphism $f : [a, b] \to [a, b]$,

$$\mathrm{CR}(f) = \mathrm{Fix}(f).$$

In fact, let $x \notin \mathrm{Fix}(f)$. We may assume $f(x) > x$. Let $\epsilon = |f(x) - x|/2$. Then there is no ϵ-chain from x to x.

Let us consider an orientation-preserving homeomorphism f that has no fixed points in (a, b). Generally, an orientation-preserving homeomorphism f restricted to (the closure of) any connected component of the complement of $\mathrm{Fix}(f)$ is such a homeomorphism. In this case, the graph of f is either above the diagonal or else below the diagonal.

Theorem 1.8. *Any two orientation-preserving homeomorphisms of $[a, b]$ without fixed points in (a, b) are topologically conjugate.*

Proof. Let f and g be two orientation-preserving homeomorphisms of $[a, b]$ without fixed points in (a, b). We assume $f(x) > x$ and $g(x) > x$ for any $x \in (a, b)$. See Figure 1.6. For the other cases the proofs are similar. Take any $p \in (a, b)$ and any homeomorphism

$$h_0 : [p, f(p)] \longrightarrow [p, g(p)]$$

such that $h_0 p = p, h_0(fp) = gp$. For each integer n, define

$$h_n : [f^n(p), f^{n+1}(p)] \longrightarrow [g^n(p), g^{n+1}(p)]$$

to be

$$h_n = g^n \circ h_0 \circ f^{-n}.$$

It is easy to see that these h_n glue together to give a homeomorphism $h : [a, b] \to [a, b]$ such that $hf = gh$. □

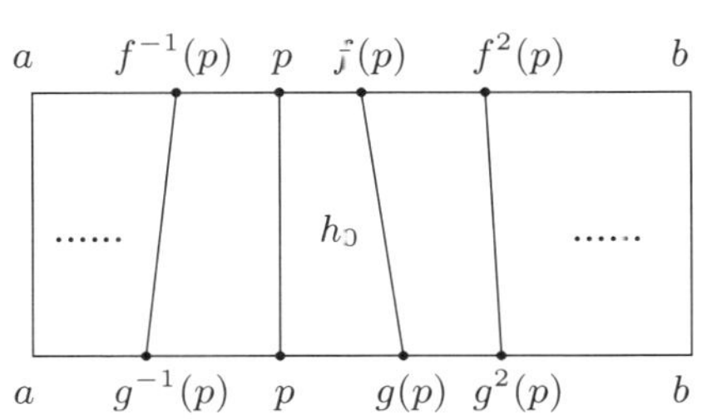

Figure 1.6. The construction of the topological conjugacy h in Theorem 1.8.

Remark. Theorem 1.8 tells us that the construction of a topological conjugacy on wandering domains is quite flexible or, from a slightly different point of view, dynamics on wandering domains are not sensitive to perturbations.

Closely related to topological conjugacy is the notion of structural stability. This is studied in the setting of differentiable dynamical systems.

Let M be a compact C^∞ manifold without boundary. Denote by $\mathrm{Diff}^r(M)$ the set of C^r diffeomorphisms of M, endowed with the C^r topology. The *C^r topology* can be defined as follows. Fix a finite cover of admissible coordinate neighborhoods (U_i, ϕ_i), $i = 1, \ldots, N$ of M. (A coordinate neighborhood (U, ϕ) is *admissible* if there is another coordinate neighborhood (V, ψ) such that $\overline{U} \subset V$ and $\phi = \psi|_U$.) We say a sequence of C^r diffeomorphisms f_n of M *converge in the C^r sense* to a C^r diffeomorphism f of M if, for all $1 \leq i, j \leq N$, the local expressions $\phi_j f_n \phi_i^{-1}$ as well as their partial derivatives up to order r converge to that of $\phi_j f \phi_i^{-1}$ uniformly at all points that make sense. This gives a C^r topology on $\mathrm{Diff}^r(M)$. Different covers of admissible coordinate neighborhoods give equivalent C^r topologies.

We say that $f \in \mathrm{Diff}^r(M)$ is *C^r structurally stable* if there is a C^r neighborhood $\mathcal{U}$ of f in $\mathrm{Diff}^r(M)$ such that every $g \in \mathcal{U}$ is topologically conjugate to f.

Briefly, f is C^r structurally stable if C^r small perturbations cannot change topologically the orbit structure of f. Here the term *perturbation* of f is not a rigorous one. It usually means a diffeomorphism g that is C^r close to f. Sometimes it also means the difference between f and g, and this is why one says "small perturbation".

Clearly, if f is C^r structurally stable, then it is C^{r+1} structurally stable. Thus C^1 structural stability is the strongest structural stability. The concept of C^0 structural stability is vacant because C^0 perturbations are too damaging. For instance, a C^0 perturbation can easily turn an isolated fixed point into a whole neighborhood of fixed points; hence it destroys any structural stability. Thus f and g have to be diffeomorphisms but not merely homeomorphisms. On the other hand, the conjugacy h must be allowed to be a homeomorphism but not necessarily a diffeomorphism. This is because a differentiable conjugacy would be too restrictive. For instance, by the chain rule, a differentiable conjugacy preserves derivatives at fixed points, while a C^r perturbation can easily change the derivative at a fixed point. Hence there would be no structurally stable systems if we require the conjugacy to be differentiable. Thus the notion of structural stability is an appropriate one. It limits to less damaging differentiable perturbations and allows a more powerful topological conjugacy to get back to the original system.

The characterization of structurally stable systems remained the central problem of differentiable dynamical systems for several decades in the last century. Nevertheless for the simplest case of an interval the problem is easy, as the next theorem shows. Note that $f \in \mathrm{Diff}^1[a, b]$ means not only that f

is a homeomorphism and differentiable, but also that f^{-1} is differentiable. (For instance $f(x) = x^3$ is not a diffeomorphism.) In particular, the absolute value of f' has a positive lower bound. It is easy to see that if f is a diffeomorphism, so is every g that is C^1 near f.

Theorem 1.9. *Let $f : [a, b] \to [a, b]$ be an orientation-preserving diffeomorphism without fixed points in (a, b). For any $r \geq 1$, f is C^r structurally stable if and only if $f'(a) \neq 1$ and $f'(b) \neq 1$.*

Proof. Assume $f'(a) \neq 1$ and $f'(b) \neq 1$. We prove f is C^r structurally stable for any $r \geq 1$. It suffices to prove it for $r = 1$. There is a C^1 neighborhood $\mathcal{U}_1$ of f in $\mathrm{Diff}^1([a, b])$, together with a neighborhood U of a and a neighborhood V of b in $[a, b]$, such that any $g \in \mathcal{U}_1$ is orientation preserving and has a unique fixed point in U which is a and a unique fixed point in V which is b. See Figure 1.7. Since f has no fixed points on $[a, b] - U - V$ which is compact, $|f(x) - x|$ assumes a positive minimum on $[a, b] - U - V$. Then there is a C^1 neighborhood $\mathcal{U}_2$ of f such that any $g \in \mathcal{U}_2$ has no fixed points on $[a, b] - U - V$. Let

$$\mathcal{U} = \mathcal{U}_1 \cap \mathcal{U}_2.$$

Then any $g \in \mathcal{U}$ is orientation preserving and has no fixed points in (a, b). By Theorem 1.8, g and f are conjugate.

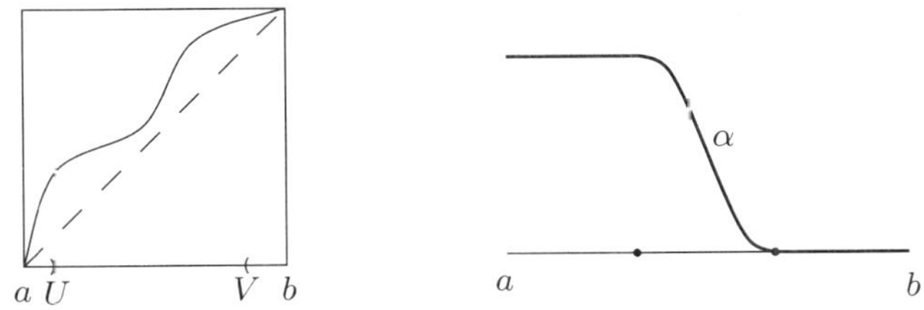

Figure 1.7. The proof of Theorem 1.9.

Conversely, assume $f'(a) = 1$ or $f'(b) = 1$. For explicitness we assume $f'(a) = 1$. For any $r \geq 1$, we construct an arbitrarily small C^r perturbation g of f such that g has more than two fixed points; hence f would not be C^r structurally stable. For simplicity we assume $a = 0$ and $b = 1$. Fix a C^∞ bump function α: $[0, 1] \to [0, 1]$ such that $\alpha = 1$ on $[0, 1/3]$, $\alpha = 0$ on $[2/3, 1]$, and $0 \leq \alpha \leq 1$ on $[0, 1]$. Without loss of generality we assume that the graph of f is above the diagonal. Then for any $\epsilon > 0$, define

$$g(x) = g_\epsilon(x) = f(x) - \epsilon\alpha(x)x.$$

If $\epsilon > 0$ is sufficiently small, g will be an orientation-preserving diffeomorphism. As $\epsilon \to 0$, g will be arbitrarily C^r close to f. Clearly, $x = 0$ and $x = 1$ are fixed points of g. Since $g(x) = f(x) - \epsilon x$ near $x = 0$, we get $g'(0) = f'(0) - \epsilon = 1 - \epsilon$. Then the graph of f is slightly rotated downward near $x = 0$ and then goes up to cross the diagonal. Hence g has another fixed point slightly to the right of $x = 0$. Thus g has at least three fixed points. This proves Theorem 1.9. □

Though being very simple, Theorem 1.9 is instructive. It indicates that, unlike the situation of wandering points, nonwandering points (a fixed point here) are sensitive to perturbations. To survive from perturbations they need a condition like $f'(a) \neq 1$, the *hyperbolicity*. Thus the nonwandering set is important not only because it absorbs the long run behavior of all orbits, but also because it is more sensitive and needs more attention from the point of view of perturbations. Another interesting point is that, as indicated by Theorem 1.9, for interval diffeomorphisms C^i structural stability is equivalent to C^j structural stability, for any i and j. This is also true for flows on orientable surfaces according to the classical result of Peixoto (1962). For general cases, this is unknown.

1.3. Circle homeomorphisms

Let S^1 be the unit circle and $f : S^1 \to S^1$ be a homeomorphism. We say f is *orientation preserving* if any lifting of f to the covering space $\mathbb{R}$ is strictly increasing, and it is *orientation reversing* if any lifting of f to the covering space $\mathbb{R}$ is strictly decreasing. Any homeomorphism $f : S^1 \to S^1$ is either orientation preserving or else orientation reversing. The composition of two orientation-preserving homeomorphisms is orientation preserving. The composition of two orientation-reversing homeomorphisms is orientation preserving. The composition of an orientation-preserving one and an orientation-reversing one is orientation reversing.

These can also be described by using oriented intervals. Any two points $a, b \in S^1$ determine two open intervals, the two connected components of $S^1 - \{a, b\}$. Denote by (a, b) the one that goes from a to b counterclockwise (here we borrow the orientation of the plane). Thus the other one is denoted (b, a), as it goes from b to a counterclockwise. Note that here the usage of the notations (a, b) and (b, a) is different from that on the real line. On the real line (a, b) and (b, a) denote the same interval with respect to different orientations of $\mathbb{R}$, while here they denote different intervals with respect to the same orientation of S^1. For any open interval $(a, b) \subset S^1$, $f(a, b)$ is an open interval with end points $f(a)$ and $f(b)$. Thus either $f(a, b) = (f(a), f(b))$ or $f(a, b) = (f(b), f(a))$. f is orientation preserving if $f(a, b) =$

$(f(a), f(b))$ for any $a, b \in S^1$, and it is orientation reversing if $f(a, b) = (f(b), f(a))$ for any $a, b \in S^1$.

Let $\Lambda \subset S^1$ be compact. By a *cointerval* of Λ we mean a connected component of $S^1 - \Lambda$. Thus a cointerval of Λ is an open interval (a, b) such that $a, b \in \Lambda$ but $(a, b) \cap \Lambda = \emptyset$. Let $f : S^1 \to S^1$ be a homeomorphism. The following lemma is straightforward, and we omit the proof.

Lemma 1.10. *Let $\Lambda \subset S^1$ be a compact invariant set of f. Then f maps cointervals to cointervals. More precisely, for any cointerval I of Λ, there is a unique cointerval J of Λ such that $f(I) = J$, and this gives a bijection on the set of cointervals of Λ.*

Example 1. Consider an orientation-preserving homeomorphism f as shown in Figure 1.8. $\text{Fix}(f)$ consists of three points. f maps each of the three cointervals of $\text{Fix}(f)$ onto itself. In each cointerval, points move in the direction that the arrows indicate. Generally, $\text{Fix}(f)$ could be any closed subset of S^1, with countable cointervals and arbitrarily assigned arrows.

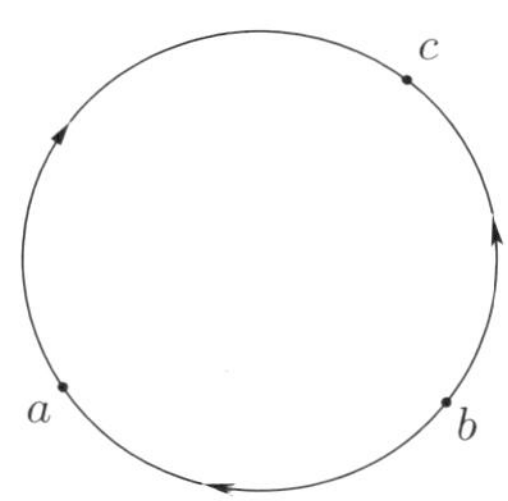

Figure 1.8. Example 1.

Example 2. A rigid rotation f of a rational angle $2\pi m/n$, $(m, n) = 1$. f is orientation preserving with $\text{P}(f) = S^1$. All periodic points have the same period n.

Example 3. A rigid rotation f of an irrational angle $2\pi\alpha$, α irrational. It is easy to see that f is orientation preserving and $\text{P}(f) = \emptyset$. We show the whole S^1 is a minimal set under f. Take any $x \in S^1$. We prove $\overline{\text{Orb}(x)} = S^1$. Take any open interval (a, b). It suffices to prove $\text{Orb}(x)$ intersects (a, b). See Figure 1.9. Since x is not periodic, $\text{Orb}(x)$ consists of infinitely many points. Hence there are two points $f^n(x)$ and $f^m(x)$ $m \neq n$, such that $l[f^n(x), f^m(x)] \leq l[a, b]/3$, where $l[a, b]$ denotes the length of the interval $[a, b]$. For simplicity denote $f^n(x) = y$, $f^m(x) = z$, and $f^{m-n} = g$. Then $g(y) = z$. Thus g is a rigid rotation of a small angle. Since $g[y, z]$, $g^2[y, z]$, $\ldots$ form a sequence of consecutive intervals of equal length $l[y, z]$, there must be $k \geq 1$ with $g^k[y, z] \subset (a, b)$. This proves that $\text{Orb}(x)$ intersects (a, b).

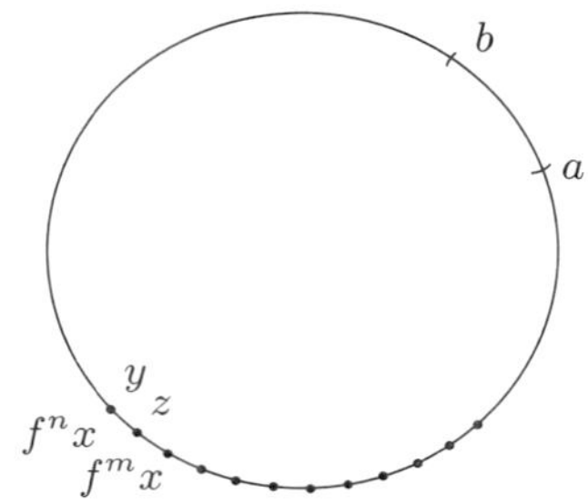

Figure 1.9. Example 3.

Example 4. An orientation-reversing homeomorphism as Figure 1.10 shows. Fix(f) consists of two fixed points $a, b \in S^1$, and f interchanges the two cointervals of $\{a, b\}$. Note that f^2 restricted to $[a, b]$, or $[b, a]$ as well, is an orientation-preserving homeomorphism. One can easily prove that $\Omega(f)$ consists of the two fixed points a and b, together with some periodic points of period 2.

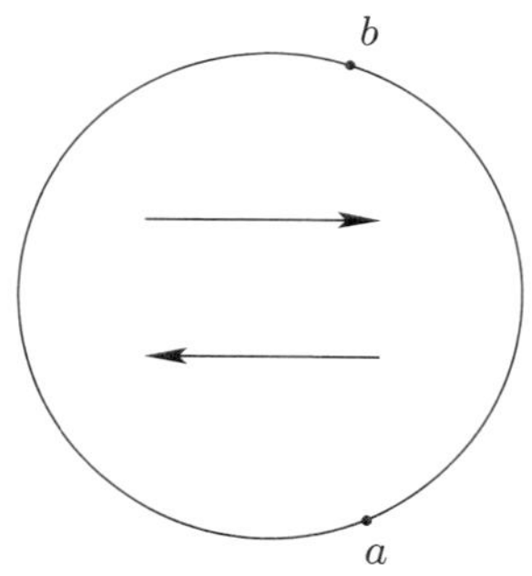

Figure 1.10. Example 4.

Since orientation-reversing homeomorphisms of S^1 have fairly simple dynamics, for the rest of the section we study orientation-preserving homeomorphisms only. We distinguish the two cases $\mathrm{P}(f) \neq \emptyset$ and $\mathrm{P}(f) = \emptyset$.

Theorem 1.11. *Let $f : S^1 \to S^1$ be an orientation-preserving homeomorphism with $\mathrm{P}(f) \neq \emptyset$. Then all periodic points of f have the same period, and $\mathrm{P}(f) = \Omega(f)$.*

Proof. Fix any $x \in \mathrm{P}(f)$. Assume the period of x is n. Let $x_1, x_2, \ldots, x_n$ be the n points of $\mathrm{Orb}(x)$, ordered counterclockwise. Note that this may not be the order of iteration. Then

$$(x_1, x_2), \ (x_2, x_3), \ \ldots, \ (x_n, x_1)$$

are the n cointervals of $\mathrm{Orb}(x)$. See Figure 1.11. Since $f^m(x_i, x_{i+1}) = (f^m x_i, f^m x_{i+1})$, if m is a multiple of n, then $f^m(x_i, x_{i+1}) = (x_i, x_{i+1})$. Otherwise $f^m(x_i, x_{i+1}) \cap (x_i, x_{i+1}) = \emptyset$.

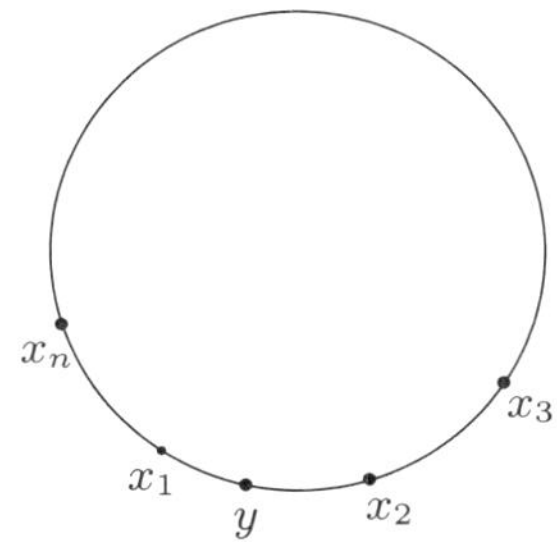

Figure 1.11. The proof of Theorem 1.11.

Take any $y \in S^1 - \mathrm{Orb}(x)$. It suffices to prove y is either n-periodic or else wandering. Without loss of generality we assume $y \in (x_1, x_2)$. Note that f^n restricted to (x_1, x_2) is an orientation-preserving homeomorphism. If $f^n(y) = y$, then y is periodic of period n. If $f^n(y) \neq y$, then there is a neighborhood V of y in (x_1, x_2) such that $f^{nk}(V) \cap V = \emptyset$ for all $k \geq 1$. As mentioned above, for any $1 \leq j \leq n-1$, $f^{nk+j}(V) \cap V = \emptyset$. Then $f^i(V) \cap V = \emptyset$ for all $i \geq 1$. Hence y is wandering. This proves Theorem 1.11. □

Now we study orientation-preserving homeomorphisms of S^1 without periodic points. Note that here we may omit the phrase "orientation preserving" because an orientation-reversing homeomorphism must have fixed points.

A *Cantor set* is defined to be a set that is compact, perfect, totally disconnected, and metrizable. See Hocking-Young (1961). All Cantor sets are mutually homeomorphic. The classical middle-third set of the interval is the best known Cantor set. Recall that a set Λ is *perfect* if it is closed and has no isolated points. In other words, every point $x \in \Lambda$ is a limit point of Λ. Also recall that a set Λ is *totally disconnected* if every connected component of Λ consists of a single point.

Theorem 1.12. *Let $f : S^1 \to S^1$ be a homeomorphism with $\mathrm{P}(f) = \emptyset$. Then $\Omega(f)$ is a minimal set. Moreover, $\Omega(f)$ either coincides with S^1 or is a Cantor set.*

Proof. To prove $\Omega(f)$ is minimal, it suffices to prove that every nonempty compact invariant set Λ of f contains $\Omega(f)$. Take any cointerval (a, b) of Λ. For any $n \geq 1$, either $f^n(a, b) = (a, b)$ or else $f^n(a, b) \cap (a, b) = \emptyset$. In the first case a and b are periodic points, contradicting $\mathrm{P}(f) = \emptyset$. Thus $f^n(a,b) \cap (a,b) = \emptyset$. This means $(a,b) \cap \Omega(f) = \emptyset$ or $\Omega(f) \subset \Lambda$.

By Theorem 1.5, either $\Omega(f) = S^1$ or $\Omega(f)$ is nowhere dense in S^1. We prove in the second case $\Omega(f)$ is a Cantor set. Note that, in dimension 1, to be nowhere dense just means to be totally disconnected. (This is not the case for dimensions higher than 1. For instance a closed interval is nowhere dense in the plane, but not totally disconnected.) Since $\Omega(f)$ is compact, it remains to prove $\Omega(f)$ contains no isolated points. Take any $x \in \Omega(f)$. Since $\Omega(f)$ is minimal, $\omega(x) = \Omega(f)$. Then there are $n_i \to \infty$ with $f^{n_i}(x) \to x$. But $\mathrm{P}(f) = \emptyset$; hence terms in the sequence $\{f^{n_i}(x)\} \subset \Omega(f)$ are mutually distinct. Thus x is not isolated in $\Omega(f)$. This proves Theorem 1.12. $\square$

A homeomorphism $f : S^1 \to S^1$ is *exceptional* if $\mathrm{P}(f) = \emptyset$ and $\Omega(f)$ is a Cantor set.

Example 5. An exceptional homeomorphism of S^1.

We informally illustrate the construction of an exceptional homeomorphism of S^1. Take a rigid irrational rotation $f : S^1 \to S^1$ and any $p \in S^1$. Replace the points

$$\dots, f^{-1}p,\ p,\ fp,\ f^2p, \dots$$

correspondingly by countably many closed intervals

$$\dots, I_{-1},\ I_0,\ I_1,\ I_2, \dots$$

with finite total length. This gives a (new) circle Σ. Define a homeomorphism $g : \Sigma \to \Sigma$ such that $gx = fx$ for all $x \in \Sigma - \bigcup_{n=-\infty}^{\infty} I_n$ and such that g maps I_n onto I_{n+1} homeomorphically in an orientation-preserving way. Then g has no periodic points but has wandering points. Hence g is an exceptional homeomorphism.

We remark that the exceptional homeomorphism g constructed in Example 5 is generally C^0 only. More elegant construction may make g a C^1 diffeomorphism. See Nitecki (1971) or Robinson (1995). But Denjoy proves that there can be no exceptional C^2 diffeomorphisms. This is a deep discovery. For references see Nitecki (1971) or de Melo and Van Strien (1993).

1.4. Conley's fundamental theorem of dynamical systems

This section introduces a result of Conley (1978) stating that every system is "gradient-like" modulo chain recurrence. It is sometimes called "the fundamental theorem of dynamical systems". It starts with a simple idea of "trapping regions" and pushes it quickly to somewhere amazingly far. We are thankful for the two unpublished notes of Franks (2001) and Bonatti (2004) on Conley theory which are very helpful to this section.

Recall that a gradient flow comes from the gradient vector field of a function $\phi : M \to \mathbb{R}$. Thus the function is strictly decreasing along the orbit of every regular point. Conley observes that every flow has a (in fact many) similar function $\phi : M \to \mathbb{R}$, called a Lyapunov function, which is constant on every chain class of the flow and strictly decreasing along the orbit of every point that is not chain recurrent. Here we state the result for systems of discrete time.

Theorem 1.13 (Conley's fundamental theorem of dynamical systems). *Let $f : X \to X$ be a homeomorphism of a compact metric space. There is a continuous function $\phi : X \to \mathbb{R}$ such that:*

(1) *If $x \notin \mathrm{CR}(f)$, then $\phi(f(x)) < \phi(x)$.*

(2) *For any $x, y \in \mathrm{CR}(f)$, $\phi(x) = \phi(y)$ if and only if x and y are in the same chain class. In particular, if $x \in \mathrm{CR}(f)$, then $\phi(f(x)) = \phi(x)$.*

(3) *$\phi(\mathrm{CR}(f))$ is a compact nowhere dense subset of $\mathbb{R}$.*

First we link chain classes to trapping regions.

An open set $U \subset X$ is called a *trapping region* of f if $f(\overline{U}) \subset U$. A compact invariant set $A \subset X$ of f is said to be *attracting* of f if it has a neighborhood U which is a trapping region such that $A = \bigcap_{n \geq 0} f^n(\overline{U})$. We call U an *isolating neighborhood* of A. A *repelling set* of f is an attracting set of f^{-1}.

If A is an attracting set of f with isolating neighborhood U, then $V = X - \overline{U}$ is a trapping region of f^{-1}; hence $A^* = \bigcap_{n \geq 0} f^{-n}(\overline{V})$ is a repelling set of f, called the *dual repelling set* of A. It is clear that A^* is independent of the choice of isolating neighborhood U for A.

Lemma 1.14. *The set of attracting sets of f is countable.*

Proof. Choose a countable basis $\mathcal{B} = \{V_n\}_{n=1}^{\infty}$ for the topology of X. Let A be an attracting set of f with isolating neighborhood U. Since A is compact, there are $V_{i_1}, \ldots, V_{i_k}$ such that $A \subset V_{i_1} \cup \cdots \cup V_{i_k} \subset U$. Then

$$A = \bigcap_{n \geq 0} f^n(U) = \bigcap_{n \geq 0} f^n(V_{i_1} \cup \cdots \cup V_{i_k}).$$

Thus there are at most as many attracting sets as finite subsets of $\mathcal{B}$, proving Lemma 1.14. □

Let $\{A_i\}_{i=1}^{\infty}$ be the attracting sets of f, and let $\{A_i^*\}$ be their dual repelling sets.

Lemma 1.15. $\mathrm{CR}(f) = \bigcap_{i=1}^{\infty}(A_i \cup A_i^*)$.

Proof. Before the proof we fix a basic fact.

Fact. For any trapping region U of f, there is ϵ_0 such that no ϵ_0-chain can go from $f^2(U)$ to $X - f(U)$.

In fact one can take $\epsilon_0 = \frac{1}{2}d(X - f(U),\ \overline{f^2(U)})$.

Now we prove "$\subset$". Suppose $x \notin A \cup A^*$ for some attracting set A. We prove $x \notin \mathrm{CR}(f)$. Let U be an isolating neighborhood of A. Note that $U - f(U)$ is a *fundamental domain* of $X - (A \cup A^*)$ in the sense that every orbit in $X - (A \cup A^*)$ meets $U - f(U)$ exactly once. Hence we may assume $x \in U - f(U)$. Take $\epsilon_1 > 0$ so that any ϵ_1-chain $x = x_0, x_1, x_2$ must have $x_2 \in f^2(U)$. Let ϵ_0 be given by the fact above. Let $\epsilon = \min\{\epsilon_0, \epsilon_1\}$. Then no ϵ-chain can go from x to x. That is, $x \notin \mathrm{CR}(f)$.

Then we prove "$\supset$". For any $x \in X$ and $\epsilon > 0$, denote by $\Omega(x, \epsilon)$ the set of $y \in X$ such that there is an ϵ-chain from x to y. Then $\Omega(x, \epsilon)$ is open. By definition, the $\epsilon/2$-neighborhood of $f(\Omega(x, \epsilon))$ is contained in $\Omega(x, \epsilon)$; hence $f(\overline{\Omega(x, \epsilon)}) \subset \Omega(x, \epsilon)$. In other words, $\Omega(x, \epsilon)$ is a trapping region.

Now let $x \notin \mathrm{CR}(f)$. Then there is an $\epsilon_0 > 0$ such that $x \notin \Omega(x, \epsilon_0)$. However, obviously, $fx \in \Omega(x, \epsilon_0)$. Let A be the attracting set with isolating neighborhood $\Omega(x, \epsilon_0)$, and let A^* be its dual repelling set. Then $x \notin A \cup A^*$. This proves Lemma 1.15. □

Remark. The last two paragraphs of the proof show that if $x \notin \mathrm{CR}(f)$, then there is an attracting set A such that x is not in A but is attracted to A. In other words, x is in the "proper basin" of A; hence it has no recurrence in any sense.

Lemma 1.16. *If $x, y \in \mathrm{CR}(f)$, then x and y are in the same chain class if and only if, for every i, x and y are either both in A_i or both in A_i^*.*

Proof. If for some i, $x \in A_i$ but $y \in A_i^*$, then there is $\epsilon_0 > 0$ such that no ϵ_0-chain can go from x to y. Hence x and y are not in the same chain class.

Conversely, if x and y are not in the same chain class, then there is $\epsilon_0 > 0$ such that either $y \notin \Omega(x, \epsilon_0)$ or $x \notin \Omega(y, \epsilon_0)$. We assume $y \notin \Omega(x, \epsilon_0)$. Let A be the attracting set with isolating neighborhood $\Omega(x, \epsilon_0)$ and A^* its dual repelling set. By Lemma 1.15, x and y are in $A \cup A^*$. Then $x \in A$, $y \in A^*$. This proves Lemma 1.16. □

Next we link trapping regions to *Lyapunov functions*, that is, functions that are nonincreasing along orbits.

Lemma 1.17. *Let (A, A^*) be an attracting-repelling pair of f. There is a continuous function $\phi : X \to [0, 1]$ such that:*

(1) $\phi|_{A^*} = 1,\ \ \phi|_A = 0,$ *and* $\phi(x) \in (0, 1)$ *for all* $x \notin A \cup A^*$.

(2) $\phi(f(x)) < \phi(x)$ *for every* $x \notin A \cup A^*$.

Proof. Let U be an isolating neighborhood of A. Take a continuous function $\alpha : X \to [0,1]$ such that $\alpha(X - U) = 1$, $\alpha(f(\overline{U})) = 0$, and $\alpha(x) \in (0,1)$ for every $x \in U - f(\overline{U})$. For instance, one can take

$$\alpha(x) = \frac{d(x, f(\overline{U}))}{d(x, X - U) + d(x, f(\overline{U}))}.$$

Let $x \notin A \cup A^*$. Then $\mathrm{Orb}(x)$ meets $U - f(U)$ exactly once; hence the values of α along $\mathrm{Orb}(x)$ is either the sequence $\ldots, 1, 1, 0, 0, \ldots$, or the sequence $\ldots 1, 1, a, 0, 0, \ldots$, where $a \in (0,1)$, depending on whether or not $\mathrm{Orb}(x)$ hits the boundary $\partial(U - f(U))$. In any case, it is a nonincreasing sequence.

Let $\alpha_n(x) = \alpha(f^n(x))$. Take a sequence $a_n > 0$ such that $\sum_{-\infty}^{\infty} a_n = 1$. Let

$$\phi = \sum_{n=-\infty}^{\infty} a_n \alpha_n.$$

As a sum of functions that are nonincreasing along orbits of f, ϕ is nonincreasing along orbits of f. We verify that ϕ satisfies the requirement. Clearly, $\phi|_{A^*} = 1$, $\phi|_A = 0$. For every $x \notin A \cup A^*$, $\alpha_n(x)$ are not constantly 1 for all n, nor constantly 0 for all n. Hence $\phi(x) \in (0,1)$. Finally, for every $x \notin A \cup A^*$, there is at least one n such that $\alpha_n(f(x)) < \alpha_n(x)$; hence $\phi(f(x)) < \phi(x)$. This proves Lemma 1.17. □

Now we finish the proof of Theorem 1.13.

Proof. By Lemma 1.17, for each $i \in \mathbb{N}$, there is $\phi_i : X \to \mathbb{R}$ with $\phi_i^{-1}(1) = A_i^*$, $\phi_i^{-1}(0) = A_i$, and ϕ_i strictly decreasing along orbits in $X - (A_i \cup A_i^*)$. Define $\phi : X \to \mathbb{R}$ by

$$\phi(x) = \sum_{i=1}^{\infty} \frac{2\phi_i(x)}{3^i}.$$

Then ϕ is continuous. If $x \notin \mathrm{CR}(f)$, then by Lemma 1.15, there is i with $x \notin A_i \cup A_i^*$. Hence $\phi(f(x)) < \phi(x)$. This proves item (1) of Theorem 1.13.

If $x \in \mathrm{CR}(f)$, then $x \in A_i \cup A_i^*$ for every i, so $\phi_i(x) = 0$ or 1 for every i. Hence the ternary expansion of $\phi(x)$ contains only the digits 0 and 2, and hence $\phi(x) \in C$, the Cantor middle-third set. Thus $\phi(\mathrm{CR}(f)) \subset C$ so $\phi(\mathrm{CR}(f))$ is compact and nowhere dense. This proves item (3) of Theorem 1.13.

Finally, if x, y are in the same chain class, then obviously $\phi(x) = \phi(y)$. We prove the converse. Let $x, y \in \mathrm{CR}(f)$ with $\phi(x) = \phi(y)$. As noticed above, $2\phi_i(x)$ is the i-th digit of the ternary expansion of $\phi(x)$; hence $\phi_i(x) = \phi_i(y)$ for every i. Then x and y are either both in A_i or both in A_i^*. By

Lemma 1.16, x and y are in the same chain class. This proves item (2), finishing the proof of Theorem 1.13. □

Exercises

Denote by X a compact metric space and by $f : X \to X$ a homeomorphism.

Exercise 1.1. Prove that $x \in X$ is periodic if and only if the orbit of x consists of finitely many points.

Exercise 1.2. Prove that $x \in X$ is periodic if and only if the orbit of x is compact.

Exercise 1.3. Construct a homeomorphism f such that $\mathrm{P}(f)$ is not closed.

Exercise 1.4. Prove that for any $x \in X$,

$$\omega(x) = \bigcap_{N \geq 0} \overline{\{f^n(x) \mid n \geq N\}}.$$

Exercise 1.5. Let $x \in X$. Prove that:

(1) $\omega(x)$ cannot be a union of two disjoint closed invariant subsets.

(2) If $\omega(x)$ is a union of finitely many periodic orbits, then $\omega(x)$ is in fact a periodic orbit.

(3) If $\omega(x)$ is a union of countably many periodic orbits, then $\omega(x)$ is in fact a periodic orbit.

How about if $\omega(x)$ is a union of uncountably many periodic orbits?

Exercise 1.6. Prove that the set $R(f)$ of recurrent points is always nonempty. Construct an example showing that $R(f)$ is not closed in X.

Exercise 1.7. Prove that $x \in X$ is nonwandering under f if and only if for any neighborhood U of x in X, there is $n \neq 0$ such that $(f^n U) \cap U \neq \emptyset$, and if and only if for any neighborhood U of x in X and any positive integer N, there is $n \geq N$ such that $(f^n U) \cap U \neq \emptyset$.

Exercise 1.8. Prove that if $\Omega(f) = X$, then $\{x : x \in \omega(x)\}$ is dense in X.

Exercise 1.9. Prove that for any $\delta > 0$ there is $\eta > 0$ such that, for any η-chain $\{x_n\}$, if a sequence of points $\{y_n\}$ satisfies $d(y_n, x_n) < \eta$, then $\{y_n\}$ is a δ-chain. In particular, a point $x \in X$ will be chain recurrent if for any $\epsilon > 0$, there is a periodic ϵ-chain going through the ϵ-neighborhood of x.

Exercise 1.10. Prove $\mathrm{CR}(f)$ is nonempty, compact, invariant and

$$\overline{\mathrm{P}(f)} \subset \mathrm{L}(f) \subset \Omega(f) \subset \mathrm{CR}(f).$$

Indicate with examples that each inclusion is strict.

Exercise 1.11. Prove that for any $\delta > 0$ there is $\eta > 0$ such that if $x_0, \ldots, x_k$ is an η-chain of f, then $x_k, \ldots, x_0$ is a δ-chain of f^{-1}.

Exercise 1.12. Prove that, in the statement of Theorem 1.4, "orbit" can be replaced by "positive orbit" or "negative orbit".

Exercise 1.13. Let $\Lambda \subset X$ be a compact invariant set of f. Prove that Λ is minimal if and only if for every $\epsilon > 0$, there is $N \geq 0$ such that for every $x \in \Lambda$, every point $y \in \Lambda$ is in the ϵ-neighborhood of the set $\{f^n(x)\}_{n=-N}^{N}$.

Exercise 1.14. Let Λ be a minimal set of f. Prove that if Λ_1 is an open invariant subset of Λ, then $\Lambda_1 = \Lambda$.

Exercise 1.15. Let $f : X \to X$ be topologically transitive. Prove that any f-invariant continuous function $\varphi : X \to R$ is constant. (φ is called *f-invariant* if $\varphi f = \varphi$.)

A compact invariant set $\Lambda \subset X$ of f is called *topologically mixing* if for any two open sets U and V of Λ, there is $N = N(U, V) \geq 1$ such that $f^n(U) \cap V \neq \emptyset$ for any $n \geq N$.

Exercise 1.16. Find a homeomorphism $f : X \to X$ that is transitive but not mixing.

Exercise 1.17. Let $f : X \to X$ and $g : X \to X$ be two homeomorphisms, and let $h : X \to X$ be a topological conjugacy from f to g. Prove that $h(\mathrm{Orb}(x, f)) = \mathrm{Orb}(h(x), g)$ for any $x \in X$. Also, prove that $h(\mathrm{P}(f)) = \mathrm{P}(g)$, $h(\Omega(f)) = \Omega(g)$, and $h(\mathrm{CR}(f)) = \mathrm{CR}(g)$.

Exercise 1.18. Prove that in the definition of C^r topology on $\mathrm{Diff}^r(M)$, different covers of admissible coordinate neighborhoods give equivalent C^r topologies.

Exercise 1.19. Give an example such that $\mathrm{CR}(f)$ cannot be decomposed into a nontrivial disjoint union of transitive sets.

Exercise 1.20. Prove that if X is connected and $\mathrm{CR}(f) = X$, then X is a chain class.

Exercise 1.21. Prove that if $f : X \to X$ has a unique chain class C, then $X = C$.

Exercise 1.22. Give an example such that $\Omega(f) \neq X$ but the unique attracting set is X.

Exercise 1.23. Let $f : S^1 \to S^1$ be an orientation-preserving homeomorphism with $\mathrm{P}(f) \neq \emptyset$. Prove that either $\mathrm{CR}(f) = \mathrm{P}(f)$ or $\mathrm{CR}(f) = S^1$.

Exercise 1.24. Let $f : S^1 \to S^1$ be an orientation-preserving homeomorphism with $\mathrm{P}(f) = \emptyset$. Prove $\mathrm{CR}(f) = S^1$.

Exercise 1.25. A continuous map $F : \mathbb{R} \to \mathbb{R}$ is called a *lifting* of $f : S^1 \to S^1$ if $\pi \circ F = f \circ \pi$, where $\pi : \mathbb{R} \to S^1$ is the projection of modulo integer parts. Let f be an orientation-preserving homeomorphism of S^1, and let F be a lifting of f. Prove:

(1) F is monotonically increasing.

(2) $F(t+1) = F(t) + 1$ for all $t \in \mathbb{R}$.

Exercise 1.26. Let $f : S^1 \to S^1$ be an orientation-preserving homeomorphism, and let $F : \mathbb{R} \to \mathbb{R}$ be a lifting of f. Prove:

(1) For any $t \in \mathbb{R}$, the limit

$$\rho(F, t) = \lim_{n\to\infty} \frac{F^n(t) - t}{n}$$

exists and is independent of t; hence it could be denoted by $\rho(F)$.

(2) If F_1 and F_2 are two liftings of f, then $\rho(F_1) - \rho(F_2)$ is an integer. Thus $\rho(f) = \rho(F) \bmod 1$ is well defined and is called the *rotation number* of f.

Exercise 1.27. Let $f, g : S^1 \to S^1$ be two orientation-preserving homeomorphisms that are topologically conjugate. Prove $\rho(f) = \rho(g)$.

Chapter 2

Hyperbolic fixed points

Hyperbolicity is the key to understanding structural stability. In this chapter we study the local behavior near a hyperbolic fixed point. The approach we take will apply directly to hyperbolic sets in Chapter 4.

2.1. Hyperbolic linear isomorphisms

Throughout this chapter, E denotes a finite-dimensional normed vector space. Two norms $|\cdot|$ and $\|\cdot\|$ of E are called *equivalent* if there is $K \geq 1$ such that, for every $v \in E$,

$$K^{-1}|v| \leq \|v\| \leq K|v|.$$

We call K a *relative constant* between these two norms. The following fact is standard in functional analysis; see for instance Bachman and Narici (1966).

Theorem 2.1. *For a finite-dimensional normed vector space all norms are equivalent.*

A linear isomorphism $A : E \to E$ is called *hyperbolic* if E splits into a direct sum

$$E = E^s \oplus E^u,$$

invariant in the sense that

$$A(E^s) = E^s, \ A(E^u) = E^u$$

such that, for two constants $C \geq 1$ and $0 < \lambda < 1$, the following uniform estimates hold:

$$|A^n v| \leq C\lambda^n |v|, \ \forall v \in E^s, \ n \geq 0,$$

$$|A^{-n} v| \leq C\lambda^n |v|, \ \forall v \in E^u, \ n \geq 0.$$

We call E^s the *contracting space* of A, and E^u the *expanding space* of A. The integer $\dim E^s$ is called the *index* of A. See Figure 2.1.

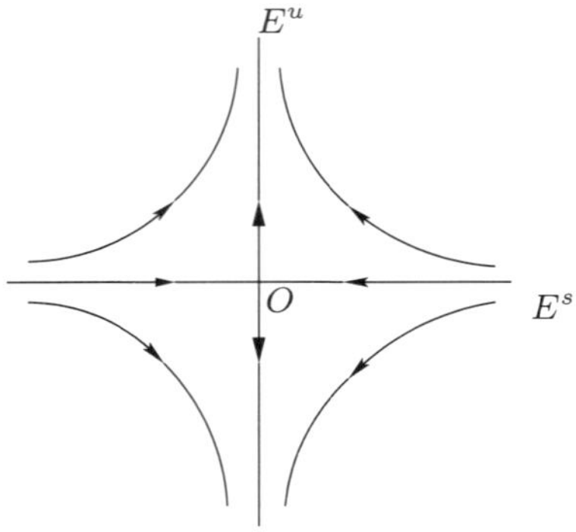

Figure 2.1. A hyperbolic linear isomorphism.

Remark. (1) Since in a finite-dimensional normed vector space all norms are equivalent, hyperbolicity is independent of the choice of norms of E.

(2) The contracting space E^s or the expanding space E^u could be $\{0\}$. In this case A is called of *source* or *sink* type. Otherwise A is said to be of *saddle* type. All results for hyperbolic linear isomorphisms in this book include the case of sink or source type, which will not be stated separately.

(3) If A is hyperbolic, so is A^{-1}.

(4) Since E^s is invariant under A, the inequalities in the definition concern not only positive but in fact all iterates of v. That is,

$$|A^n(A^m v)| \leq C\lambda^n |A^m v|, \ \forall v \in E^s, \ m \in \mathbb{Z}, \ n \geq 0.$$

Note that m and n play different roles here: m represents the moment when the iterate starts, while n represents the stride it crosses. It is n that serves as the exponent of λ. See Figure 2.2. In particular, letting $m = -n$ gives

$$|A^{-n}v| \geq C^{-1}(\lambda^{-1})^n |v|, \ \forall v \in E^s, \ n \geq 0.$$

Likewise for E^u.

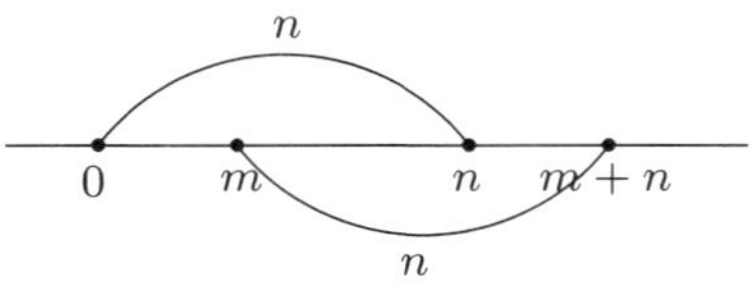

Figure 2.2. An iterate of stride n.

A simple and equivalent definition for a linear isomorphism A to be hyperbolic is that all the eigenvalues of A have norms different from 1. The definition given above is more complicated, but it applies to the general case of a hyperbolic set of Chapter 4. Thus we will study hyperbolic linear isomorphisms according to the above definition so that the results and methods of this chapter can be extended to Chapter 4 directly.

For $\gamma > 0$, denote by

$$C_\gamma(E^s) = \{v \in E \mid |v_u| \le \gamma |v_s|\},$$

$$C_\gamma(E^u) = \{v \in E \mid |v_s| \le \gamma |v_u|\}$$

the *γ-cones* about E^s and E^u, respectively. Here v_s and v_u denote the components of v in E^s and E^u, respectively.

Theorem 2.2 (Characterization of E^s). *Let $A : E \to E$ be a hyperbolic linear isomorphism with splitting $E = E^s \oplus E^u$. Then E^s is characterized by*

$$\begin{aligned} E^s &= \{v \in E \mid A^n v \to 0, n \to +\infty\} \\ &= \{v \in E \mid \text{there is } r > 0 \text{ such that } |A^n v| \le r, \forall n \ge 0\} \\ &= \{v \in E \mid \text{there is } \gamma > 0 \text{ such that } A^n v \in C_\gamma(E^s), \forall n \ge 0\}. \end{aligned}$$

Likewise,

$$\begin{aligned} E^u &= \{v \in E \mid A^{-n} v \to 0, n \to +\infty\} \\ &= \{v \in E \mid \text{there is } r > 0 \text{ such that } |A^{-n} v| \le r, \forall n \ge 0\} \\ &= \{v \in E \mid \text{there is } \gamma > 0 \text{ such that } A^{-n} v \in C_\gamma(E^u), \forall n \ge 0\}. \end{aligned}$$

In particular, hyperbolic splitting is unique. That is, if $E = G^s \oplus G^u$ is another hyperbolic splitting of A, then $G^s = E^s$, $G^u = E^u$.

Proof. We prove this for E^s. Obviously the first set is contained in the second, and the second is contained in the third. We prove that the third is contained in the fourth. In fact, if

$$v \notin \{v \in E \mid \text{there is } \gamma > 0 \text{ such that } A^n v \in C_\gamma(E^s), \forall n \ge 0\},$$

then there is $m \ge 0$ such that $w = A^m v \notin C_1(E^s)$. In particular, $w_u \ne 0$. Then

$$|A^n(w_u)| \to \infty, \quad |A^n(w_s)| \to 0$$

as $n \to +\infty$. Hence

$$|A^n w| \ge |A^n(w_u)| - |A^n(w_s)| \to \infty.$$

Hence $\{|A^n v|\}_{n=0}^{+\infty}$ is unbounded.

Finally we prove

$$\{v \in E \mid \text{there is } \gamma > 0 \text{ such that } A^n v \in C_\gamma(E^s), \forall n \ge 0\} \subset E^s.$$

In fact, if $v \notin E^s$, then $v_u \neq 0$. Then

$$|A^n(v_u)| \to \infty, \quad |A^n(v_s)| \to 0$$

as $n \to +\infty$. Thus for any $\gamma > 0$, $A^n v \notin C_\gamma(E^s)$ for n large.

The proof for E^u is similar. Since E^s and E^u have been characterized, the uniqueness of hyperbolic splitting follows automatically. □

Thus, upon iteration of a hyperbolic linear isomorphism, every nonzero vector exhibits extremal dynamical behavior: It either approaches 0 exponentially fast or diverges to infinity exponentially fast.

Theorem 2.3. *Let A be a hyperbolic linear isomorphism with splitting $E = E^s \oplus E^u$. There are a norm $\|\cdot\|$ of E and a constant $0 < \tau < 1$ such that*

$$\|Av\| \le \tau\|v\|, \ \forall v \in E^s,$$

$$\|A^{-1}v\| \le \tau\|v\|, \ \forall v \in E^u.$$

Briefly, there is a norm that makes the hyperbolic behavior an immediate contraction and expansion.

Proof. Let $|\cdot|$ be the original norm of E. Take N sufficiently large such that $C\lambda^N < 1$, and define

$$\|v\| = \sum_{n=0}^{N-1} |A^n v|, \ \forall\ v \in E.$$

Then $\|\cdot\|$ is a norm of E. Let $a = \sum_{n=0}^{N-1} C\lambda^n$. Then

$$\|v\| \le a|v|, \ \forall\ v \in E^s,$$

$$\|v\| \le a|A^{N-1}v|, \ \forall\ v \in E^u.$$

We check that $\|\cdot\|$ satisfies the two inequalities of the theorem. In fact, for any $v \in E^s$,

$$\begin{aligned}\|Av\| &= \|v\| - |v| + |A^N v| \\ &\le \|v\| - (1 - C\lambda^N)|v| \le \|v\| - a^{-1}(1 - C\lambda^N)\|v\|.\end{aligned}$$

Likewise, for any $v \in E^u$,

$$\begin{aligned}\|A^{-1}v\| &= \|v\| + |A^{-1}v| - |A^{N-1}v| \\ &\le \|v\| - (1 - C\lambda^N)|A^{N-1}v| \le \|v\| - a^{-1}(1 - C\lambda^N)\|v\|.\end{aligned}$$

Let

$$\tau = 1 - a^{-1}(1 - C\lambda^N).$$

It suffices to verify $0 < \tau < 1$. But this is obvious because $a \ge 1$. □

A norm $\|\cdot\|$ of E that satisfies the two immediate inequalities of Theorem 2.3 is said to be *adapted* to A. We call

$$\tau(A) = \max\{\|A|_{E^s}\|,\ \|A^{-1}|_{E^u}\|\} < 1$$

the *skewness* of A with respect to this adapted norm.

Let $E = E_1 \oplus E_2$ be a direct sum. A norm $|\cdot|$ of E is said to be of *box type* with respect to $E_1 \oplus E_2$ if

$$|v| = \max\{|v_1|, |v_2|\},\ \forall\ v \in E,$$

where v_1 and v_2 are the two components of v with respect to $E_1 \oplus E_2$. If a norm is of box type with respect to the hyperbolic splitting $E^s \oplus E^u$ of a hyperbolic linear isomorphism A, we often simply say that the norm is of box type with respect to A.

For any norm $|\cdot|$ of E, letting

$$\|v\| = \max\{|v_1|, |v_2|\},\ \forall\ v \in E,$$

where v_1 and v_2 are the two components of v with respect to $E_1 \oplus E_2$, defines a norm $\|\cdot\|$ of E that is of box type with respect to $E_1 \oplus E_2$, called the *box-adjusted norm* of $|\cdot|$ with respect to $E_1 \oplus E_2$. Clearly, the box-adjusted norm of an adapted norm to a hyperbolic linear isomorphism A with respect to its hyperbolic splitting is adapted to A with the same skewness; hence it is both adapted to and of box type to A.

2.2. Persistence of hyperbolic fixed points

Let $O \subset E$ be an open set, and let $f : O \to E$ be a C^1 map. A fixed point $p \in O$ of f is called *hyperbolic* if the derivative $Df(p) : E \to E$ is a hyperbolic linear isomorphism. The *index* of p is defined to be the index of $Df(p)$.

The main aim of this chapter is to investigate the behavior of f near a hyperbolic fixed point p. By the inverse function theorem, f near p is a diffeomorphism onto its image. Hence there is an open set U with $p \in U \subset \overline{U} \subset O$, $\overline{U}$ compact, such that $f : U \to E$ is a diffeomorphism onto its image. The main theorems of this chapter will deal with such a triple f, p, U, starting with stating the following: Let $p \in U$ be a hyperbolic fixed point of f....

Let $f,\ g : U \to E$ be two C^r maps. Define the C^r *distance* between f and g to be

$$d^r(f,g) = \sup_{x\in U}\{|f(x) - g(x)|,\ |Df(x) - Dg(x)|, \ldots, |D^r f(x) - D^r g(x)|\}.$$

Here the norm of a higher-order derivative is just the operator norm of its associated multilinear operator. Note that we are using $|\cdot|$ to denote the operator norm. The notation $\|\cdot\|$ will be used for a norm of E that is different from $|\cdot|$. We will mainly use the C^1 metric.

Denote by $\mathcal{B}^1(f,\delta) = \mathcal{B}^1(f,\delta;U)$ the set of diffeomorphisms $g : U \to E$ onto its image satisfying $d^1(g,f) \le \delta$. As usual, denote by $B(p,r)$ the closed ball of center p and radius $r > 0$. If $\|\cdot\|$ is another norm of E, then for any $\delta > 0$, there is $\eta > 0$ such that $\mathcal{B}^1(f,\eta;\|\cdot\|) \subset \mathcal{B}^1(f,\delta;|\cdot|)$. Likewise for $B(p,r)$.

Recall that a map $\phi : E \to E$ is called *Lipschitz* if there is a constant $k \ge 0$ such that

$$|\phi x - \phi y| \le k|x - y|, \quad \forall\, x, y \in E.$$

The minimal number k that satisfies the inequality is called the *Lipschitz constant* of ϕ, denoted $\operatorname{Lip}\phi$.

While the main concern of this text is about C^1 perturbations, many results hold for more general Lipschitz perturbations, and the treatment is even easier. In this case, we will work out the proof in the Lipschitz setting as a lemma. The C^1 statement itself will follow directly as a consequence. The following classical result, for instance see Rudin (1976), serves as the bridge.

The generalized mean value theorem. *Let $B \subset E$ be a convex open set, and let $f : B \to E$ be a C^1 map such that $|Df(x)| \le k$ for any $x \in B$. Then for any $x, y \in B$, $|f(x) - f(y)| \le k|x - y|$.*

That is, on a convex open set, the Lipschitz constant is bounded above by the supremum of the norms of the derivatives. A direct consequence is

Lemma 2.4. *Let $f : U \to E$ be a C^1 map, and let $p \in U$ be a point. For any $\epsilon > 0$, there are $\delta > 0$ and $r > 0$ such that, for any $g \in \mathcal{B}^1(f,\delta)$, $\operatorname{Lip}(g - Df(p)) \le \epsilon$ on $B(p,r)$.*

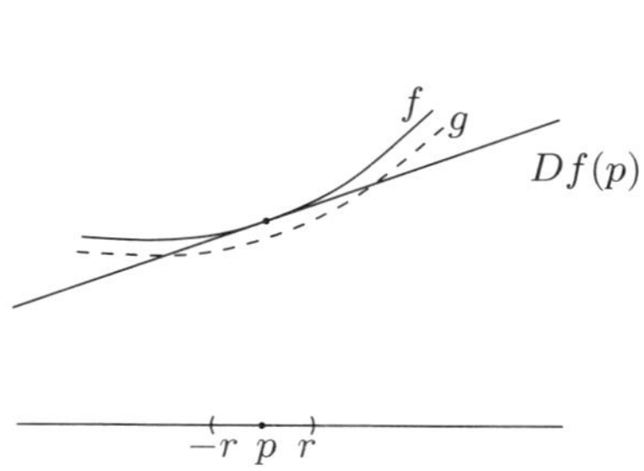

Figure 2.3. The statement of Lemma 2.4.

Proof. Let $\epsilon > 0$ be given. There are $\delta > 0$ and $r > 0$ such that for any $g \in \mathcal{B}^1(f,\delta)$ and any $x \in B(p,r)$,

$$|D(g - Df(p))(x)| = |Dg(x) - Df(p)| \leq \epsilon.$$

Then Lemma 2.4 follows directly from the generalized mean value theorem. □

Let $E = E^s \oplus E^u$ be a direct sum. Denote by

$$\pi_s : E \to E^s \quad \text{and} \quad \pi_u : E \to E^u$$

the two projections. For any map $\phi : E \to E$, denote

$$\phi_s = \pi_s \circ \phi \quad \text{and} \quad \phi_u = \pi_u \circ \phi.$$

If $A : E \to E$ is linear, denote

$$A_{ss} = A_s|_{E^s} \quad \text{and} \quad A_{uu} = A_u|_{E^u}.$$

It is easy to see that if E^s and E^u are A-invariant, then for any $v \in E$,

$$A_s v = A_s v_s = A_{ss} v_s,$$
$$A_u v = A_u v_u = A_{uu} v_u.$$

For simplicity denote by

$$E(r) = \{v \in E \mid |v| \leq r\}$$

the closed r-ball about the origin; that is, $E(r) = B(0,r)$.

By Lemma 2.4, on a neighborhood of a hyperbolic fixed point $p \in U$, one may write f as well as its C^1 perturbation g in the form

$$A + \phi,$$

where A is a hyperbolic linear isomorphism and $\operatorname{Lip}\phi$ is small. This will be the form we use frequently, which could be referred to as a *Lipschitz perturbation of a hyperbolic linear isomorphism*.

Lemma 2.5. *Let $A : E \to E$ be a hyperbolic linear isomorphism with splitting $E = E^s \oplus E^u$, and let $|\cdot|$ be a norm of E that is adapted to and of box type to A. Let $0 < \tau < 1$ be the skewness of A with respect to $|\cdot|$. Let $r > 0$. If $\phi : E(r) \to E$ is Lipschitz with*

$$\operatorname{Lip}\phi < 1 - \tau,$$

then $A + \phi$ has in $E(r)$ at most one fixed point. If, in addition,

$$|\phi(0)| \leq (1 - \tau - \operatorname{Lip}\phi)r,$$

then $A+\phi$ has in $E(r)$ at least one (hence a unique) fixed point p_ϕ. Moreover,

$$|p_\phi| \leq \frac{|\phi(0)|}{1 - \tau - \operatorname{Lip}\phi}.$$

Proof. We solve the equation

$$(A + \phi)v = v$$

for $v \in E(r)$. Writing in components with respect to the direct sum

$$E = E^s \oplus E^u,$$

this is equivalent to solving

$$A_s v + \phi_s v = v_s, \; A_u v + \phi_u v = v_u,$$

or

$$A_{ss} v_s + \phi_s v = v_s, \; A_{uu} v_u + \phi_u v = v_u,$$

or

$$A_{ss} v_s + \phi_s v = v_s, \; A_{uu}^{-1} v_u - A_{uu}^{-1} \phi_u v = v_u.$$

This suggests a map

$$T = T_\phi : E(r) \to E$$

$$T(v) = (A_{ss} v_s + \phi_s v, \; A_{uu}^{-1} v_u - A_{uu}^{-1} \phi_u v).$$

Then T and $A + \phi$ have the same set of fixed points. Therefore, to prove that $A + \phi$ has at most one fixed point in $E(r)$, it suffices to prove T is a contraction. Since

$$\begin{aligned} &|T(v) - T(v')| \\ &\quad \le \max\{\tau |v_s - v'_s| + \operatorname{Lip}\phi \cdot |v - v'|, \; \tau |v_u - v'_u| + \tau \operatorname{Lip}\phi \cdot |v - v'|\} \\ &\quad \le (\tau + \operatorname{Lip}\phi)|v - v'|, \end{aligned}$$

and since

$$\operatorname{Lip}\phi < 1 - \tau,$$

T is indeed a contraction. This proves that $A + \phi$ has in $E(r)$ at most one fixed point.

Now assume in addition that

$$|\phi(0)| \le (1 - \tau - \operatorname{Lip}\phi) r.$$

We prove T maps $E(r)$ into itself. Take any $v \in E(r)$. Since

$$|T(0)| = |(\phi_s(0), \; -A_{uu}^{-1} \phi_u(0))| \le |\phi(0)|,$$

we have

$$\begin{aligned} |T(v)| &\le |T(0)| + |T(v) - T(0)| \\ &\le |\phi(0)| + (\tau + \operatorname{Lip}\phi)|v| \le r; \end{aligned}$$

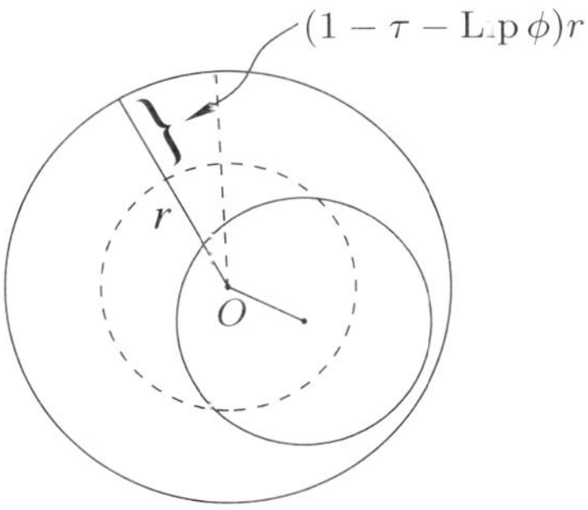

Figure 2.4. The proof of Lemma 2.5. The smaller circle is of center $T(0)$ and of radius at most $(\tau + \operatorname{Lip}\phi)r$; hence it must be contained in the bigger circle.

see Figure 2.4. Thus T maps $E(r)$ into itself. By the contraction mapping principle, T has a (unique) fixed point p_ϕ in $E(r)$. Moreover, letting $v = p_\phi$ in the above inequality

$$|T(v)| \leq |\phi(0)| + (\tau + \operatorname{Lip}\phi)|v|$$

gives

$$|p_\phi| \leq |\phi(0)| + (\tau + \operatorname{Lip}\phi)|p_\phi|.$$

Hence

$$|p_\phi| \leq \frac{|\phi(0)|}{1-\tau-\operatorname{Lip}\phi}.$$

This proves Lemma 2.5. □

The following theorem is an application of Lemma 2.5 to differentiable maps.

Theorem 2.6 (Persistence of hyperbolic fixed point). *Let $p \in U$ be a hyperbolic fixed point of f. There are $\delta_0 > 0$ and $\epsilon_0 > 0$ such that any $g \in \mathcal{B}^1(f, \delta_0)$ has in $B(p, \epsilon_0)$ at most one fixed point. Moreover, for any $0 < \epsilon \leq \epsilon_0$, there is $0 < \delta \leq \delta_0$ such that any $g \in \mathcal{B}^1(f, \delta)$ has in $B(p, \epsilon)$ at least one (hence unique) fixed point p_g.*

Remark. Briefly, $p_g \to p$ when g C^1 approaches f. That is, p_g varies continuously in g.

Proof. If the statement holds for one norm of E, it holds for every norm. It hence suffices to prove the theorem under a norm $|\cdot|$ of E that is adapted to and of box type to $Df(p)$. Without loss of generality we assume $p = 0$. Abbreviate $Df(0) = A$. Let $0 < \tau < 1$ be the skewness of A with respect to this norm. Fix

$$\tau < \lambda < 1.$$

By Lemma 2.4, there are $\delta_0 > 0$ and $\epsilon_0 > 0$ such that, for any $g \in \mathcal{B}^1(f, \delta_0)$,

$$\phi = g - A : E(\epsilon_0) \to E$$

satisfies

$$\operatorname{Lip} \phi \leq \lambda - \tau$$

on $E(\epsilon_0)$. By Lemma 2.5, $g = A + \phi$ has in $E(\epsilon_0)$ at most one fixed point.

Let $0 < \epsilon \leq \epsilon_0$ be given. Let

$$\delta = \min\{\delta_0, (1 - \lambda)\epsilon\}.$$

For any $g \in \mathcal{B}^1(f, \delta)$,

$$\begin{aligned} |\phi(0)| &= |g(0)| = |g(0) - f(0)| \\ &\leq d^1(g, f) \leq (1 - \lambda)\epsilon. \end{aligned}$$

By Lemma 2.5, g has in $E(\epsilon_0)$ at least one (hence unique) fixed point p_g, and

$$|p_g| \leq \frac{(1 - \lambda)\epsilon}{1 - \tau - \operatorname{Lip} \phi_g} \leq \epsilon.$$

This proves Theorem 2.6. □

The unique fixed point p_g of g in $B(p, \epsilon_0)$ is called the *continuation* of p under g. This notion is defined for g sufficiently C^1 close to f.

2.3. Persistence of hyperbolicity for a fixed point

Hyperbolicity is persistent under perturbations. That is, linear maps that are near a hyperbolic linear isomorphism A are hyperbolic linear isomorphisms. Note that the definition of hyperbolicity involves not only contraction and expansion, but also invertibility. Thus we start with recalling the linear version of the inverse function theorem stating that linear maps that are near an invertible one are invertible. For more general goals we will need the following Lipschitz inverse function theorem stating that Lipschitz maps that are near an invertible linear map are invertible. We state it for the more general case when the domain E and the range E' are different.

Let $(E, |\cdot|)$ and $(E', |\cdot|')$ be two finite-dimensional normed vector spaces of the same dimension. We call a homeomorphism $f : E \to E'$ a *lipeomorphism* if both f and f^{-1} are Lipschitz. For a linear map $A : E \to E'$, we call

$$m(A) = \inf\{|Av|' \mid v \in E, \ |v| = 1\}$$

the *mininorm* of A. If A is invertible, then $m(A) = |A^{-1}|^{-1}$.

Theorem 2.7 (Lipschitz inverse function theorem). *Let $A : E \to E'$ be a linear isomorphism, and let $\phi : E \to E'$ be Lipschitz. If*

$$\operatorname{Lip}\phi < m(A),$$

then $A + \phi : E \to E'$ is a lipeomorphism and

$$\operatorname{Lip}((A+\phi)^{-1}) \leq \frac{1}{m(A) - \operatorname{Lip}\phi}.$$

Proof. First we prove $A + \phi$ is 1-1 and onto. This means that, for any $z \in E'$, the equation

$$(A+\phi)x = z$$

has a unique solution for $x \in E$, or

$$x = A^{-1}z - A^{-1}\phi(x)$$

has a unique solution for $x \in E$. In other words, we need to prove the map

$$T : E \to E$$
$$T(x) = A^{-1}z - A^{-1}\phi(x)$$

has a unique fixed point. By the contraction mapping principle, it suffices to check that T is a contraction. For any $x, y \in E$,

$$\begin{aligned} |Tx - Ty| &= |A^{-1}\phi x - A^{-1}\phi y| \\ &\leq |A^{-1}| \cdot \operatorname{Lip}\phi \cdot |x - y|. \end{aligned}$$

By assumption $|A^{-1}| \cdot \operatorname{Lip}\phi < 1$. Thus T is indeed a contraction. This proves that $A + \phi$ is 1-1 and onto.

Clearly $A + \phi$ is Lipschitz. We check that $(A+\phi)^{-1}$ is Lipschitz with the specified constant. For any $x, y \in E$,

$$\begin{aligned} |(A+\phi)x - (A+\phi)y| &\geq |A(x-y)| - |\phi x - \phi y| \\ &\geq (m(A) - \operatorname{Lip}\phi)|x - y|. \end{aligned}$$

Treating $x = (A+\phi)^{-1}x'$ and $y = (A+\phi)^{-1}y'$ gives

$$\operatorname{Lip}((A+\phi)^{-1}) \leq \frac{1}{m(A) - \operatorname{Lip}\phi}.$$

This proves Theorem 2.7. □

Remark. A special case is when ϕ is C^1. Combining the above with the inverse function theorem, it is easy to see that $A + \phi : E \to E'$ is a diffeomorphism. An even more special case is when ϕ is linear. In this case

$\operatorname{Lip}\phi$ is just the operator norm of ϕ, and the theorem just concludes that $A + \phi : E \to E'$ is a linear isomorphism.

Let A and X be metric spaces. Below for the product space $A \times X$ we will always use the "box metric" defined by

$$d((a,x),\ (b,y)) = \max\{d(a,b),\ d(x,y)\}.$$

Here we have used the same letter d to denote the metrics in the three spaces A, X, and $A \times X$. This will not cause confusion in general. The next theorem concerns how the fixed points depend on parameters for a uniform family of contractions. See Figure 2.5.

Theorem 2.8. *Let A and X be two metric spaces with X complete, and let $F : A \times X \to X$ be a map. Assume there is $0 < \lambda < 1$ such that*

$$d(F(a,x), F(a,y)) \le \lambda d(x,y)$$

for any $a \in A$ and any $x, y \in X$. Denote by $p(a)$ the (unique) fixed point of $F(a,\cdot)$. This gives a map $p : A \to X$. Then:

(1) *If F is continuous, so is p.*

(2) *If F is Lipschitz, so is p.*

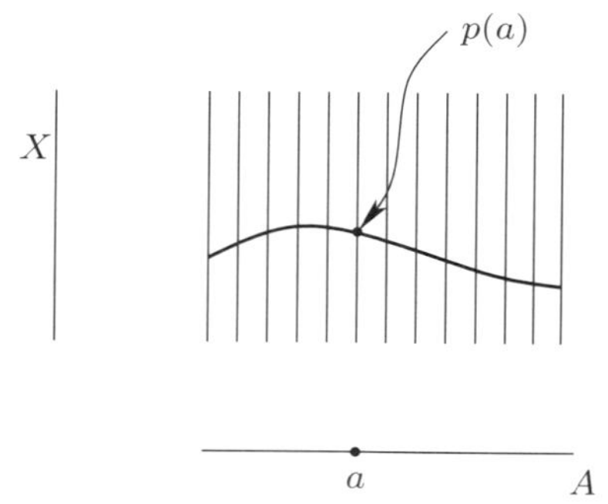

Figure 2.5. The statement of Theorem 2.8. Each vertical line is a copy of X, contracted into itself by F; hence it has a unique fixed point, denoted by $p(a)$ for the copy of parameter value a.

Proof. Since

$$\begin{aligned} d(p(a), p(b)) &= d(F(a,p(a)),\ F(b,p(b))) \\ &\le d(F(a,p(a)),\ F(a,p(b))) + d(F(a,p(b)),\ F(b,p(b))) \\ &\le \lambda d(p(a),p(b)) + d(F(a,p(b)),\ F(b,p(b))), \end{aligned}$$

it follows that

$$d(p(a),p(b)) \le \frac{1}{1-\lambda} d(F(a,p(b)),\ F(b,p(b))).$$

Then (1) and (2) follow immediately. This proves Theorem 2.8. □

Now we prove that linear maps B that are near a hyperbolic isomorphism A are hyperbolic. Let $E^s \oplus E^u$ be the hyperbolic splitting of A. We prove that B has a hyperbolic splitting $G^s \oplus G^u$. See Figure 2.6.

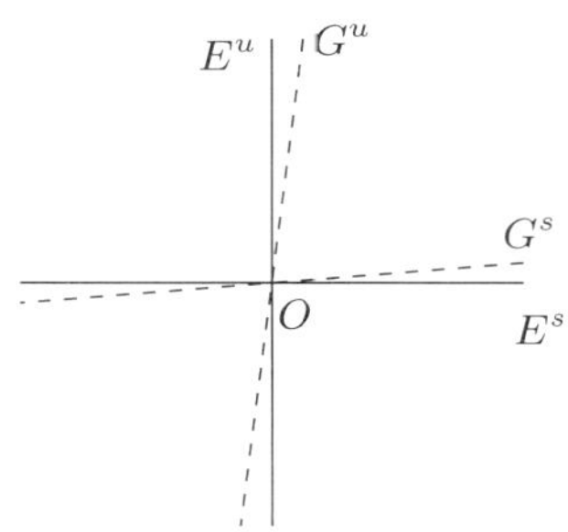

Figure 2.6. The splittings of A and B, respectively.

Let us find G^u. (G^s can be found in the same way.) The idea is very intuitive. In fact, A contracts the (opening of the) unstable cones $C_1(E^u)$, and E^u is just the "kernel"

$$\bigcap_{n=0}^{\infty} A^n(C_1(E^u))$$

of the contraction. Since B contracts the unstable cones $C_1(E^u)$ as well, this suggests that the set G^u we are looking for should be

$$\bigcap_{n=0}^{\infty} B^n(C_1(E^u));$$

see Figure 2.7. In particular, we need to show that this intersection is a linear subspace of E. This idea can be realized more directly by considering the set of graphs of linear maps from E^u to E^s of norm ≤ 1. Every such graph is already a linear subspace of E that lies in $C_1(E^u)$. Intuitively, B transforms these graphs into one another. The one we are looking for, G^u, is first of all invariant under B, which amounts to a fixed point of the "graph transform". The fact that B contracts the (opening of the) unstable cones just means it contracts the (angle)-distance between any two such graphs. Thus we are led to a problem of the contraction mapping principle. The next lemma carries out this idea. It skips A and refers to B directly.

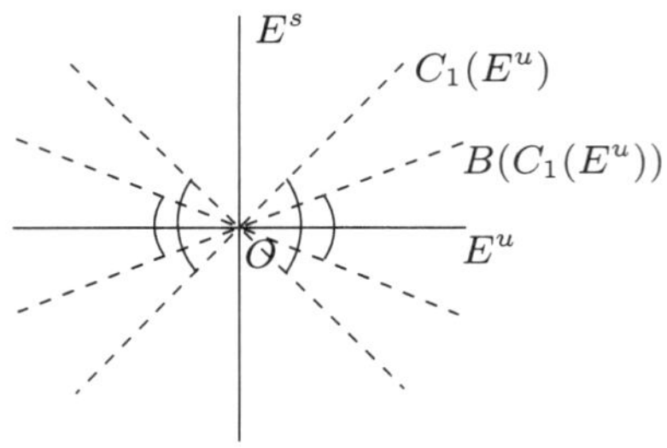

Figure 2.7. Contraction of cones.

Let $E = E_1 \oplus E_2$ be a direct sum. Denote by $L(E_1, E_2)$ the space of linear maps from E_1 to E_2 and by $L(E_1, E_2)(1)$ the closed unit ball about the origin.

Lemma 2.9. *Let $B : E \to E$ be a linear isomorphism, represented under a direct sum $E = E_1 \oplus E_2$ as*

$$\begin{pmatrix} B_{11} & B_{12} \\ B_{21} & B_{22} \end{pmatrix},$$

where $B_{ij} = \pi_i \circ B|_{E_j}$. If there are a norm $|\cdot|$ of E that is of box type with respect to $E_1 \oplus E_2$ and two constants $\lambda > 0$ and $\epsilon > 0$ such that

$$\max\{|B_{11}^{-1}|, |B_{22}|\} < \lambda,$$
$$\max\{|B_{12}|, |B_{21}|\} < \epsilon,$$
$$\lambda + \epsilon < 1,$$

then there is a unique linear map $P = P_B : E_1 \to E_2$, $|P| \leq 1$, such that the linear subspace $\mathrm{gr}(P)$ is B-invariant and $B|_{\mathrm{gr}(P)}$ is $(\lambda^{-1} - \epsilon)$-expanding. Moreover, P_B, and hence $\mathrm{gr}(P_B)$, depends continuously on B.

Remark. $\lambda + \epsilon < 1$ implies $\lambda^{-1} - \epsilon > 1$.

Proof. Let $P : E_1 \to E_2$ be a linear map with $|P| \leq 1$. For any $v \in E_1$,

$$\begin{pmatrix} B_{11} & B_{12} \\ B_{21} & B_{22} \end{pmatrix} \begin{pmatrix} v \\ Pv \end{pmatrix} = \begin{pmatrix} B_{11}v + B_{12}Pv \\ B_{21}v + B_{22}Pv \end{pmatrix}.$$

Hence

$$B(\mathrm{gr}(P)) \subset \mathrm{gr}(P)$$

if and only if

$$P(B_{11}v + B_{12}Pv) = B_{21}v + B_{22}Pv, \ \forall v \in E_1;$$

that is,

$$P(B_{11} + B_{12}P) = B_{21} + B_{22}P.$$

Since

$$m(B_{11}) \geq \lambda^{-1}, \quad |B_{12}P| \leq \epsilon,$$

by Theorem 2.7,

$$B_{11} + B_{12}P : E_1 \to E_1$$

is invertible. Hence

$$P = (B_{21} + B_{22}P)(B_{11} + B_{12}P)^{-1}.$$

This suggests a map

$$T = T_B : L(E_1, E_2)(1) \to L(E_1, E_2),$$

$$T(P) = (B_{21} + B_{22}P)(B_{11} + B_{12}P)^{-1},$$

naturally called the *graph transform* induced by B. Finding a linear map P with $B(\mathrm{gr}(P)) \subset \mathrm{gr}(P)$ then reduces to finding a fixed point of T.

We verify that T maps $L(E_1, E_2)(1)$ into itself and is a contraction. In fact, for any $P \in L(E_1, E_2)(1)$,

$$\begin{aligned}|T(P)| &\leq |B_{21} + B_{22}P| \cdot |(B_{11} + B_{12}P)^{-1}| \\ &\leq \frac{\lambda + \epsilon}{\lambda^{-1} - \epsilon} < 1.\end{aligned}$$

Hence T maps $L(E_1, E_2)(1)$ into itself. Moreover, for any $P, P' \in L(E_1, E_2)(1)$,

$$T(P)(B_{11} + B_{12}P) = B_{21} + B_{22}P,$$

$$T(P')(B_{11} + B_{12}P') = B_{21} + B_{22}P'.$$

Hence

$$\begin{aligned}(T(P) - T(P'))B_{11} + T(P)B_{12}P - T(P')B_{12}P + T(P')B_{12}P - T(P')B_{12}P' \\ = B_{22}(P - P'),\end{aligned}$$

$$(T(P) - T(P'))(B_{11} + B_{12}P) = (B_{22} - T(P')B_{12})(P - P'),$$

$$T(P) - T(P') = (B_{22} - T(P')B_{12})(P - P')(B_{11} + B_{12}P)^{-1}.$$

Hence

$$|T(P) - T(P')| \leq \frac{\lambda + \epsilon}{\lambda^{-1} - \epsilon}|P - P'|.$$

Thus $T = T_B$ is a contraction. By the contraction mapping principle, T has a unique fixed point $P = P_B \in L(E_1, E_2)(1)$. In other words, there is a unique $P = P_B \in L(E_1, E_2)(1)$ such that $B(\mathrm{gr}(P)) \subset \mathrm{gr}(P)$. Since $B : E \to E$ is a linear isomorphism, the inclusion is actually an equality

$$B(\mathrm{gr}(P)) = \mathrm{gr}(P).$$

Since the norm is of box type and since $P \in L(E_1, E_2)(1)$, the norm of a vector in $\mathrm{gr}(P)$ is given by the first component. Then

$$|B(v, Pv)| = |B_{11}v + B_{12}Pv| \geq (\lambda^{-1} - \epsilon)|v|;$$

that is, $B|_{\mathrm{gr}(P)}$ is $(\lambda^{-1} - \epsilon)$-expanding.

Let $\mathcal{B}$ denote the set of linear isomorphisms that satisfy the assumptions of Lemma 2.9. What is discussed above is a family of contractions

$$T : \mathcal{B} \times L(E_1, E_2)(1) \to L(E_1, E_2)(1)$$

$$T(B, P) = T_B(P) = (B_{21} + B_{22}P)(B_{11} + B_{12}P)^{-1}$$

with parameter B. Clearly T is continuous. As the above computation shows, the contraction rate of T_B is independent of $B \in \mathcal{B}$. By Theorem 2.8, the fixed point P_B, hence the graph $\mathrm{gr}(P_B)$, varies continuously in B. This proves Lemma 2.9. □

Theorem 2.10 (Persistence of hyperbolicity for a linear map). *Let $A : E \to E$ be a hyperbolic linear isomorphism. There is $\delta_0 > 0$ such that if a linear map $B : E \to E$ satisfies $|B - A| < \delta_0$, then B is hyperbolic. Moreover, the stable and unstable spaces $E^s(B)$ and $E^u(B)$ vary continuously in B.*

Proof. If the theorem holds for one norm of E, it will hold for every norm of E. It hence suffices to prove the theorem under a norm $|\cdot|$ of E that is adapted to and of box type to A. Let $E = E^u \oplus E^s$ be the hyperbolic splitting of $A : E \to E$ with skewness $0 < \tau < 1$ with respect to $|\cdot|$. Then

$$|A_{uu}^{-1}| \leq \tau, \quad |A_{ss}| \leq \tau, \quad A_{us} = A_{su} = 0.$$

By Theorem 2.7 (applied for the case when ϕ is linear), there is $\delta_0 > 0$ such that if a linear map $B : E \to E$ satisfies $|B - A| < \delta_0$, then B is invertible. Choose $\tau < \lambda < 1$ and $\epsilon > 0$ such that $\lambda + \epsilon < 1$. Shrink $\delta_0 > 0$ so that if $|B - A| < \delta_0$, then B satisfies Lemma 2.9 with respect to the direct sum $E^u \oplus E^s$. By the identity

$$A^{-1} - B^{-1} = A^{-1}(B - A)B^{-1},$$

if B is near A, then B^{-1} is near A^{-1}. Shrink $\delta_0 > 0$ further so that if $|B - A| < \delta_0$, then B^{-1} satisfies Lemma 2.9 with respect to $E^s \oplus E^u$ as well. Hence there are $P_B \in L(E^u, E^s)(1)$ and $Q_B \in L(E^s, E^u)(1)$ such that $G^u = \mathrm{gr}(P_B)$ and $G^s = \mathrm{gr}(Q_B)$ are both invariant under B and are expanding and contracting, respectively. In particular, $G^u \cap G^s = \{0\}$. Since the two linear subspaces have complementary dimensions, it follows that $E = G^u \oplus G^s$. Thus B is hyperbolic with $E^s(B) = G^s$ and $E^u(B) = G^u$. The continuity is stated in Lemma 2.9. This proves Theorem 2.10. □

Combining Theorems 2.6 and 2.10 gives

Theorem 2.11. *Let $p \in U$ be a hyperbolic fixed point of f. Then there are $\delta_0 > 0$ and $\epsilon_0 > 0$ such that every $g \in \mathcal{B}^1(f, \delta_0)$ has in $B(p, \epsilon_0)$ a unique fixed point p_g, which is hyperbolic. Moreover, p_g, as well as $E^s(p_g)$ and $E^u(p_g)$, varies continuously with g.*

A periodic point $p \in U$ of f of period m is called *hyperbolic* if the derivative $Df^m(p) : E \to E$ is a hyperbolic linear isomorphism. Problems on hyperbolic periodic points often reduce to problems on hyperbolic fixed points.

2.4. Hartman-Grobman theorem

The Hartman-Grobman theorem asserts that a diffeomorphism f restricted to a neighborhood of a hyperbolic fixed point $p \in E$ is topologically conjugate to its derivative $Df(p)$ restricted to a neighborhood of the origin. We postpone the specifications of the two neighborhoods in question and first consider the ideal setting of the whole space E. Let $C^0(E)$ be the set of continuous maps from E to itself. Denote

$$C_b^0(E) = \left\{ \phi \in C^0(E) \mid \sup_{x \in E} |\phi(x)| < \infty \right\},$$

endowed with the C^0 norm

$$|\phi| = \sup_{x \in E} |\phi(x)|.$$

This is a Banach space. The following lemma of Pugh (1969) is a Lipschitz version of the Hartman-Grobman theorem.

Lemma 2.12. *Let $A : E \to E$ be a hyperbolic linear isomorphism of skewness $0 < \tau < 1$ with respect to a box norm $|\cdot|$ of E that is adapted to and of box type to A. Assume $\phi,\ \psi \in C_b^0(E)$ satisfy*

$$\max\{\operatorname{Lip} \phi,\ \operatorname{Lip} \psi\} < \min\{1 - \tau,\ m(A)\}.$$

Then there is a unique $\eta \in C_b^0(E)$ such that $id + \eta : E \to E$ is a homeomorphism with

$$(id + \eta)(A + \phi) = (A + \psi)(id + \eta).$$

Proof. We solve the equation

$$(*) \qquad (id+\eta)(A+\phi) = (A+\psi)(id+\eta)$$

for $\eta \in C_b^0(E)$. This is equivalent to

$$A + \phi + \eta(A + \phi) = A + A\eta + \psi(id + \eta),$$

or

$$\phi + \eta(A + \phi) = A\eta + \psi(id + \eta).$$

Writing in components with respect to the direct sum

$$E = E^s \oplus E^u,$$

we get

$$\phi_s + \eta_s(A + \phi) = A_{ss}\eta_s + \psi_s(id + \eta), \quad \phi_u + \eta_u(A + \phi) = A_{uu}\eta_u + \psi_u(id + \eta),$$

or

$$\eta_s = (A_{ss}\eta_s + \psi_s(id + \eta) - \phi_s)(A + \phi)^{-1}, \quad \eta_u = A_{uu}^{-1}(\phi_u + \eta_u(A + \phi) - \psi_u(id + \eta)).$$

Note that $A + \phi$ is indeed invertible by Theorem 2.7. This suggests a map

$$T : C_b^0(E) \to C_b^0(E)$$

$$T(\eta) = ((A_{ss}\eta_s + \psi_s(id + \eta) - \phi_s)(A + \phi)^{-1}, \ A_{uu}^{-1}(\phi_u + \eta_u(A + \phi) - \psi_u(id + \eta))).$$

Note that since $\phi, \psi, \eta \in C_b^0(E)$, $T(\eta)$ is indeed in $C_b^0(E)$. Clearly, a solution of $(*)$ is just a fixed point of T. We check that T is a contraction mapping. In fact,

$$\begin{aligned}
&|T_s(\eta) - T_s(\eta')| \\
&\quad = |(A_{ss}(\eta_s - \eta'_s) + \psi_s(id + \eta) - \psi_s(id + \eta'))(A + \phi)^{-1}| \\
&\quad = \sup_{x \in E} |(A_{ss}(\eta_s - \eta'_s) + \psi_s(id + \eta) - \psi_s(id + \eta'))(A + \phi)^{-1}(x)| \\
&\quad = \sup_{y \in E} |(A_{ss}(\eta_s - \eta'_s) + \psi_s(id + \eta) - \psi_s(id + \eta'))(y)| \\
&\quad \le \sup_{y \in E} (\tau \cdot |\eta_s(y) - \eta'_s(y)| + \operatorname{Lip} \psi \cdot |\eta(y) - \eta'(y)|) \\
&\quad \le (\tau + \operatorname{Lip} \psi)|\eta - \eta'|.
\end{aligned}$$

Likewise,

$$\begin{aligned}
|T_u(\eta) - T_u(\eta')| &\le (\tau + \tau \operatorname{Lip} \psi)|\eta - \eta'| \\
&\le (\tau + \operatorname{Lip} \psi)|\eta - \eta'|.
\end{aligned}$$

This verifies that T is a contraction mapping; hence it has a unique fixed point $\eta \in C_b^0(E)$. That means there is a unique $\eta \in C_b^0(E)$ such that

$$(id + \eta)(A + \phi) = (A + \psi)(id + \eta).$$

Interchanging ϕ and ψ gives a unique $\xi \in C_b^0(E)$ such that

$$(id + \xi)(A + \psi) = (A + \phi)(id + \xi).$$

The two commutative diagrams in Figure 2.8 combined together give a self-conjugacy

$$(id + \xi)(id + \eta)(A + \phi) = (A + \phi)(id + \xi)(id + \eta)$$

for $A+\phi$, where $(id+\xi)(id+\eta)$ is clearly in the class $id+C_b^0$. See Figure 2.8. Since id is already a self-conjugacy for $A+\phi$ in the class $id+C_b^0$, by uniqueness,

$$(id+\xi)(id+\eta)=id.$$

Likewise

$$(id+\eta)(id+\xi)=id.$$

Thus $id+\eta$ is a homeomorphism. This proves Lemma 2.12. $\square$

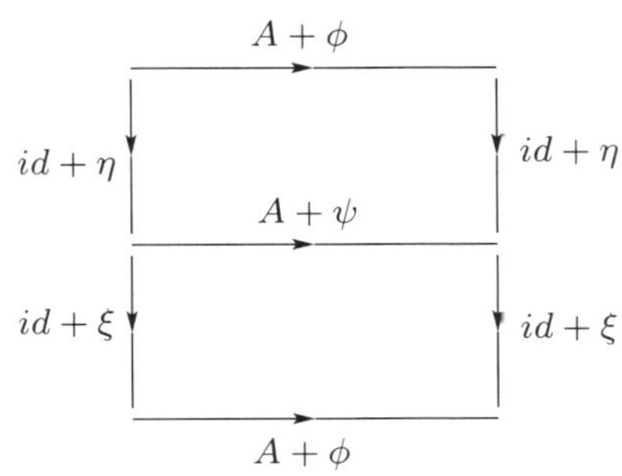

Figure 2.8. The conjugacy from $A+\phi$ to $A+\psi$ composed with the conjugacy from $A+\psi$ to $A+\phi$ gives a conjugacy from $A+\phi$ to itself.

Theorem 2.13 (Hartman-Grobman). *Let $0\in U$ be a hyperbolic fixed point of f. There are a neighborhood V of 0 in E and a homeomorphism $h: V\cup f(V)\to E$ onto its image such that $h\circ f|_V = Df(0)\circ h|_V$.*

Note that h is not a topological conjugacy in a strict sense because neither $f|_V$ nor $Df(0)|_{hV}$ is a dynamical system.

Proof. If the theorem holds for one norm of E, then it holds for every norm of E. It hence suffices to prove the theorem under a norm $|\cdot|$ of E that is adapted to and of box type to $Df(0)$. Let $0<\tau<1$ be the skewness of $Df(0)$. Fix a C^∞ bump function $\alpha: E\to\mathbb{R}$ with $0\le\alpha\le 1$ such that $\alpha(x)=1$ for $|x|\le 1/3$ and $\alpha(x)=0$ for $|x|\ge 2/3$. Take $B>0$ such that $|D\alpha(x)|\le B$ for all $x\in E$.

Write $A=Df(0)$. Define

$$\phi = f-A: U\to E.$$

Then

$$\phi(0)=0,\quad D\phi(0)=0.$$

Define

$$\overline{\phi}: E\to E$$
$$\overline{\phi}(x)=\alpha(x/r)\phi(x),$$

where $r > 0$ satisfies

$$E(r) \subset U$$

and is to be reduced further. Then $\overline{\phi} \in C_b^0(E)$ is C^1, which coincides with ϕ on $E(r/3)$ and vanishes outside $E(2r/3)$. See Figure 2.9.

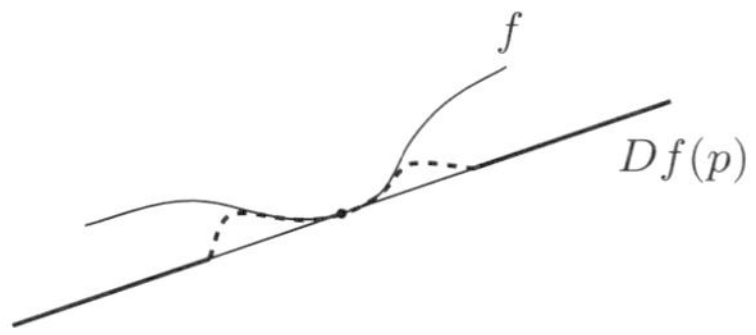

Figure 2.9. The map $A + \overline{\phi}$.

Claim. $\operatorname{Lip} \overline{\phi} \to 0$ as $r \to 0$.

By the generalized mean value theorem, it suffices to prove $\sup_{x \in E}|D\overline{\phi}(x)| \to 0$ as $r \to 0$. Taking derivative we have

$$|D\overline{\phi}(x)| \leq |D\alpha(x/r)| \cdot \frac{1}{r} \cdot |\phi(x)| + |\alpha(x/r)| \cdot |D\phi(x)|.$$

Since

$$\phi(0) = 0, \quad D\phi(0) = 0,$$

$\phi(x)$ is $o(|x|)$. For any $\epsilon > 0$, there is $\delta > 0$ such that if $|x| \leq \delta$, then $|\phi(x)| \leq \frac{\epsilon}{2B}|x|$ and $|D\phi(x)| \leq \epsilon/2$. Since $\overline{\phi}$ vanishes outside $E(r)$, to compute $|D\overline{\phi}(x)|$ we may assume $|x| \leq r$. Hence if $r < \delta$, then

$$\begin{aligned} |D\overline{\phi}(x)| &\leq B \cdot \frac{1}{r} \cdot \frac{\epsilon}{2B}|x| + 1 \cdot \frac{\epsilon}{2} \\ &\leq \epsilon/2 + \epsilon/2 = \epsilon. \end{aligned}$$

This proves the claim.

Thus, if $r > 0$ is sufficiently small, we will have

$$\operatorname{Lip} \overline{\phi} < \min\{1 - \tau, \ m(A)\}$$

on the whole E. By Lemma 2.12, there is a homeomorphism $h : E \to E$ such that

$$h \circ (A + \overline{\phi}) = A \circ h.$$

Let $V = E(r/3)$. Then

$$h \circ f|_V = Df(0) \circ h|_V.$$

This proves Theorem 2.13. □

The Harman-Grobman theorem is perhaps the only result of Chapter 2 that will not generalize to the hyperbolic sets of Chapter 4. We include it here for its obvious importance and elegance.

2.5. The local stable manifold for a hyperbolic fixed point

Let $p \in U$ be a hyperbolic fixed point of f. For $r > 0$, define the *local stable manifold* and *local unstable manifold* of p of size r to be respectively,

$$W_r^s(p, f) = \{v \in U \mid |f^n v - p| \leq r \ \forall n \geq 0, \text{ and } \lim_{n \to +\infty} f^n v = p\},$$

$$W_r^u(p, f) = \{v \in U \mid |f^{-n} v - p| \leq r \ \forall n \geq 0, \text{ and } \lim_{n \to +\infty} f^{-n} v = p\}.$$

Of course here r has to satisfy $B(p, r) \subset U$. Clearly,

$$f(W_r^s(p)) \subset W_r^s(p), \quad f(W_r^u(p)) \supset W_r^u(p).$$

There are several equivalent characterizations for the local stable manifold. Up to a translation we assume $p = 0$.

Lemma 2.14 (Characterization of W_r^s, adapted form). *Let $A : E \to E$ be a hyperbolic linear isomorphism with splitting $E = E^s \oplus E^u$ of skewness $0 < \tau < 1$ with respect to a norm $|\cdot|$ of E that is adapted to and of box type to A. Let $r > 0$. Let $\phi : E(r) \to E$ be Lipschitz such that*

$$\operatorname{Lip} \phi < 1 - \tau, \qquad \phi(0) = 0.$$

Then

$$\begin{aligned} W_r^s(0, A + \phi) &= \{v \in E(r) \mid |(A + \phi)^n v| \leq r, \ \forall n \geq 0\} \\ &= \{v \in E(r) \mid (A + \phi)^n v \in E(r) \cap C_1(E^s), \ \forall n \geq 0\} \\ &= \{v \in E(r) \mid |(A + \phi)^n v| \leq (\tau + \operatorname{Lip} \phi)^n |v|, \ \forall n \geq 0\}. \end{aligned}$$

Likewise for W_r^u.

Proof. First we prove two simple facts.

Claim 1. *If $v, v' \in E(r)$, then*

$$|(A + \phi)_s v - (A + \phi)_s v'| \leq (\tau + \operatorname{Lip} \phi)|v - v'|.$$

In fact,

$$\begin{aligned} |(A + \phi)_s v - (A + \phi)_s v'| &= |A_{ss}(v_s - v_s') + \phi_s(v) - \phi_s(v')| \\ &\leq (\tau + \operatorname{Lip} \phi)|v - v'|. \end{aligned}$$

Claim 2. *If $v, v' \in E(r)$ and $v - v' \notin C_1(E^s)$, then*

$$(A + \phi)v - (A + \phi)v' \notin C_1(E^s)$$

and

$$|(A + \phi)_u v - (A + \phi)_u v'| \geq (\tau^{-1} - \operatorname{Lip} \phi)|v - v'|.$$

(Note that $\tau^{-1} - \operatorname{Lip} \phi > 1$.)

Briefly, Claim 2 says that if two points of $E(r)$ are in "vertical position", so will their images be. Moreover, the distance gets expanded.

In fact,

$$\begin{aligned}|(A+\phi)_u v-(A+\phi)_u v'| &= |A_{uu}(v_u-v'_u)+\phi_u(v)-\phi_u(v')|\\ &\geq \tau^{-1}|v_u-v'_u|-\operatorname{Lip}\phi|v-v'|.\end{aligned}$$

But $v-v'\notin C_1(E^s)$; hence $|v_u-v'_u|=|v-v'|$. Then

$$|(A+\phi)_u v-(A+\phi)_u v'|\geq(\tau^{-1}-\operatorname{Lip}\phi)|v-v'|.$$

Now $v-v'\notin C_1(E^s)$; hence $v-v'\neq 0$. Combining this with Claim 1 we get

$$|(A+\phi)_u v-(A+\phi)_u v'|>|(A+\phi)_s v-(A+\phi)_s v'|.$$

Thus $(A+\phi)v-(A+\phi)v'\notin C_1(E^s)$. This proves Claim 2.

We prove the equivalent conditions circularly. Obviously, the first set is contained in the second. We prove that the second is contained in the third. We use the two claims for the special case $v'=0$ (the general case will be used in the proof of Lemma 2.17). Assume there is $v\in E(r)$ such that $(A+\phi)^n v\in E(r)$ for all $n\geq 0$ but there is $m\geq 0$ such that $w=(A+\phi)^m v\notin C_1(E^s)$. By Claim 2,

$$|(A+\phi)w|\geq(\tau^{-1}-\operatorname{Lip}\phi)|w|,$$

and $(A+\phi)w\notin C_1(E^s)$. Inductively, for any $n\geq 1$,

$$|(A+\phi)^n w|\geq(\tau^{-1}-\operatorname{Lip}\phi)^n|w|.$$

Note that $w\neq 0$ as $w\notin C_1(E^s)$. Thus $\{(A+\phi)^n w\}_{n=0}^{\infty}$ is unbounded, a contradiction.

Next we prove that the third set is contained in the forth. Assume for any $n\geq 0$, $(A+\phi)^n v\in E(r)\cap C_1(E^s)$. By Claim 1,

$$|(A+\phi)v|=|(A+\phi)_s v|\leq(\tau+\operatorname{Lip}\phi)|v|.$$

Inductively, for any $n\geq 1$,

$$|(A+\phi)^n v|\leq(\tau+\operatorname{Lip}\phi)^n|v|.$$

Finally, it is obvious that the forth set is contained in the first. This proves Lemma 2.14. □

Remark. In Lemma 2.14, $E(r)$ could be the whole E. In this case the formulation of the lemma should be slightly changed; for instance the second set $\{v\in E(r)\mid(A+\phi)^n v\in E(r),\ \forall n\geq 0\}$ should of course be changed to $\{v\in E\mid \text{there is } r>0 \text{ such that } (A+\phi)^n v\in E(r),\ \forall n\geq 0\}$. The proof is the same as the case when r is fixed and hence is omitted.

We apply Lemma 2.14 to our diffeomorphism f.

Theorem 2.15 (Characterizations of W_r^s, general form). *Let $p \in U$ be a hyperbolic fixed point of f. There are $r > 0$, $C \geq 1$, and $0 < \lambda < 1$ such that*

$$W_r^s(p,f) = \{v \in U \mid |f^n v - p| \leq r,\ \forall\, n \geq 0\}$$
$$= \{v \in U \mid |f^n v - p| \leq r, \text{ and } |f^n v - p| \leq C\lambda^n |v - p|,\ \forall\, n \geq 0\}.$$

Likewise,

$$W_r^u(p,f) = \{v \in U \mid |f^{-n} v - p| \leq r,\ \forall\, n \geq 0\}$$
$$= \{v \in U \mid |f^{-n} v - p| \leq r,\ C\, |f^{-n} v - p| \leq C\lambda^n |v - p|,\ \forall\, n \geq 0\}.$$

Proof. We prove the theorem for W_r^s only. If the statement holds for one norm of E, it holds for every norm. It hence suffices to prove the theorem under a norm $|\cdot|$ of E that is adapted to and of box type to $Df(p)$. It suffices to prove that there are $r > 0$, $C \geq 1$, and $0 < \lambda < 1$ such that the second set is contained in the third.

Without loss of generality we assume $p = 0$. Write $Df(0) = A$. Let $0 < \tau < 1$ be the skewness of A with respect to $|\cdot|$. Let $C = 1$, and fix

$$\tau < \lambda < 1.$$

By Lemma 2.4, there is $r > 0$ sufficiently small such that

$$\phi = f - A : E(r) \to E$$

satisfies on $E(r)$

$$\operatorname{Lip}\phi \leq \lambda - \tau.$$

Let

$$|f^n v| \leq r,\ \forall\, n \geq 0.$$

Then

$$|f^n v| = |(A + \phi)^n v| \leq (\tau + \operatorname{Lip}\phi)^n |v| \leq \lambda^n |v|,$$

where the first inequality is by Lemma 2.14. This proves Theorem 2.15. □

Theorem 2.16 (Isolation of a hyperbolic fixed point). *Let $p \in U$ be a hyperbolic fixed point of f. There is $r > 0$ such that if $w \in U$ satisfies*

$$|f^n w - p| \leq r, \quad \forall\, n \in \mathbb{Z},$$

then $w = p$.

Proof. Take $r > 0$, $C \geq 1$, and $0 < \lambda < 1$ such that Theorem 2.15 holds for both f and f^{-1}. Let $w \in U$ satisfy $|f^n w - p| \leq r$ for all $n \in \mathbb{Z}$. By Theorem 2.15, for any $n \geq 0$,

$$|w - p| = |f^{-n}(f^n w) - p| \leq C\lambda^n |f^n w - p| \leq C^2 \lambda^{2n} |w - p|.$$

Here the first "$\leq$" holds because the $f^n w$ are within r of p for all negative iterates, and the second "$\leq$" holds because the w are within r of p for all positive iterates. Taking n sufficiently large then gives $w = p$. □

Now we proceed to the main part of the local stable manifold theorem which states that, for a hyperbolic fixed point $p \in U$, the local stable manifold $W_r^s(p)$ of p is a differentiable submanifold of the domain U, as smooth as the map f, tangent at p to the stable summand E^s. Note that by the Hartman-Grobman theorem we know that $W_r^s(p)$ is a topological submanifold of U.

The problem of the stable manifold has a long history. For historical notes see Hartman (1964) and Robinson (1995). There are two basic types of proofs for the stable manifold theorem, the graph transform method of Hadamard and the variation of parameters method of Perron. The proof given below uses the graph transform method and is influenced by Hirsh-Pugh-Shub (1977), Katok-Hasselblatt (1995), and Robinson (1995).

As we did in the proof of the Hartman-Grobman theorem, we will first leave aside the choice of neighborhoods and work out the proof of a lemma on the whole space E under a norm that is adapted to and of box type to A.

Thus we consider the ideal setting $A+\phi : E \to E$, where A is a hyperbolic linear isomorphism, $\phi(0) = 0$, and $\operatorname{Lip}\phi$ is small on the whole E. Define the (global) *unstable manifold* of $0 \in E$ to be

$$W^u(0, A+\phi) = \left\{ v \in E \mid \lim_{n\to+\infty} (A+\phi)^{-n} v = 0 \right\}.$$

The next lemma states that in this case $W^u(0, A+\phi)$ will be a Lipschitz copy of E^u. Moreover, if ϕ is C^1, so will $W^u(0, A+\phi)$ be.

Lemma 2.17. *Let $A : E \to E$ be a hyperbolic linear isomorphism with splitting $E = E^u \oplus E^s$, and let $|\cdot|$ be a norm of E that is adapted to and of box type to A. Then there is $\delta > 0$ such that:*

(1) *If $\phi : E \to E$ is Lipschitz such that*

$$\operatorname{Lip}\phi < \delta, \quad \phi(0) = 0,$$

then there is a Lipschitz map $\sigma : E^u \to E^s$, $\sigma(0) = 0$, $\operatorname{Lip}\sigma \le 1$, such that $W^u(0, A+\phi) = \operatorname{gr}(\sigma)$.

(2) *If $\phi : E \to E$ is C^1 such that*

$$\operatorname{Lip}\phi < \delta, \quad \phi(0) = 0,$$

then the map $\sigma : E^u \to E^s$ guaranteed by item (1) is C^1 and the C^1 submanifold $W^u(0, A+\phi)$ of E is tangent at the origin to the unstable subspace G^u of the hyperbolic linear isomorphism $A + D\phi(0)$.

Remark. If $\operatorname{Lip}\phi$ is small, then $A + D\phi(0)$ and $(A + D\phi(0))^{-1}$ will both satisfy the assumptions of Lemma 2.9; hence $A + D\phi(0)$ will be hyperbolic.

Proof. We first prove item (1). Let $0 < \tau < 1$ be the skewness of A with respect to $|\cdot|$. Let

$$\delta = \min\left\{\frac{1-\tau}{2},\ m(A)\right\}.$$

(Later we will reduce δ further.)

Let $\phi : E \to E$ be a Lipschitz map such that

$$\mathrm{Lip}(\phi) < \delta, \quad \phi(0) = 0.$$

We prove there is a Lipschitz map $\sigma : E^u \to E^s$, $\sigma(0) = 0$, $\mathrm{Lip}\,\sigma \leq 1$, whose graph $\mathrm{gr}(\sigma)$ is invariant under $A + \phi$. Then we prove that the graph is exactly $W^u(0, A + \phi)$.

The invariance condition

$$(A + \phi)(\mathrm{gr}(\sigma)) \subset \mathrm{gr}(\sigma)$$

is equivalent to

$$\sigma((A + \phi)_u(v + \sigma v)) = (A + \phi)_s(v + \sigma v)$$

for every $v \in E^u$. Since $A_u(\sigma v) = 0$ and $A_s v = 0$, this reduces to

$$\sigma(A_{uu}v + \phi_u(v + \sigma v)) = A_{ss}(\sigma v) + \phi_s(v + \sigma v).$$

That is,

$$\sigma(A_{uu} + \phi_u(I_u + \sigma)) = A_{ss}\sigma + \phi_s(I_u + \sigma).$$

Since

$$m(A_{uu}) \geq \tau^{-1}, \quad \mathrm{Lip}(\phi_u(I_u + \sigma)) \leq 2\mathrm{Lip}\,\phi < 2\delta = 1 - \tau,$$

by Theorem 2.7, $A_{uu} + \phi_u(I_u + \sigma)$ is invertible. Hence

$$\sigma = (A_{ss}\sigma + \phi_s(I_u + \sigma))(A_{uu} + \phi_u(I_u + \sigma))^{-1}.$$

This suggests a map

$$T(\sigma) = (A_{ss}\sigma + \phi_s(I_u + \sigma))(A_{uu} + \phi_u(I_u + \sigma))^{-1},$$

called the *graph transform* induced by $A + \phi$. See Figure 2.10. Finding σ with

$$(A + \phi)(\mathrm{gr}(\sigma)) \subset \mathrm{gr}(\sigma)$$

then reduces to finding a fixed point of T.

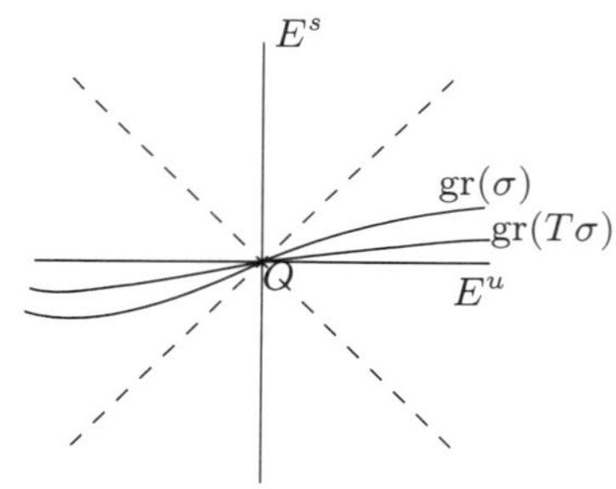

Figure 2.10. $A+\phi$ transforms the graph $\mathrm{gr}(\sigma)$ to the graph $\mathrm{gr}(T\sigma)$, inducing a map T that transforms σ to $T(\sigma)$. We are to find the invariant graph under $A+\phi$, which corresponds to the fixed point of T.

We wish to find an appropriate domain for T so that T is a contraction. This is delicate. Denote

$$\Sigma(E^u, E^s; 0) = \{\sigma : E^u \to E^s \mid \sigma(0) = 0,\ |\sigma|_* < \infty\}$$

where

$$|\sigma|_* = \sup_{v \neq 0} \frac{|\sigma(v)|}{|v|}.$$

If σ is linear, it is just the usual operator norm. With this norm $\Sigma(E^u, E^s; 0)$ forms a Banach space, and

$$\Sigma(E^u, E^s; 0)[1] = \{\sigma \in \Sigma(E^u, E^s; 0) \mid \sigma \text{ is Lipshits and } \mathrm{Lip}\,\sigma \leq 1\}$$

is a closed subset of $\Sigma(E^u, E^s; 0)$. It is a proper subset of the unit ball $\Sigma(E^u, E^s; 0)(1)$. The graph transform is then defined to be

$$T = T_\phi : \Sigma(E^u, E^s; 0)[1] \to \Sigma(E^u, E^s; 0),$$

$$T(\sigma) = (A_{ss}\sigma + \phi_s(I_u + \sigma))(A_{uu} + \phi_u(I_u + \sigma))^{-1}.$$

Remark. To prove that the invariant graph is C^1, it would be natural to use the C^1 norm for some set of C^1 maps in question and to show that the graph transform is a contraction of this set, with respect to the C^1 norm. There are counterexamples, however, showing that the graph transform is not a contraction with respect to the C^1 norm. This is a delicate and serious difficulty in the proof of the stable manifold theorem. We will first choose an appropriate norm so that the graph transform is a contraction with a Lipschitz invariant graph; then we prove that the graph is actually C^1 using different ideas, as the second step. Even in the first step, it is delicate to choose the space and the norm, as well as the domain for the contraction. Here we have followed Hirsch-Pugh-Shub (1977) to look at the space $\Sigma(E^u, E^s; 0)$ with the norm $|\cdot|_*$ and to choose the subset $\Sigma(E^u, E^s; 0)[1]$ to be the domain.

We verify T maps $\Sigma(E^u, E^s; 0)[1]$ into itself. Since $\sigma \in \Sigma(E^u, E^s; 0)[1]$, it is easy to see that $(T\sigma)(0) = 0$ and $(T\sigma)$ is Lipschitz with

$$\mathrm{Lip}(T\sigma) \leq \frac{\tau + 2\mathrm{Lip}\,\phi}{\tau^{-1} - 2\mathrm{Lip}\,\phi} < 1.$$

Hence T maps $\Sigma(E^u, E^s; 0)[1]$ into itself.

Next we verify that T is a contraction with respect to the norm $|\cdot|_*$. For any $\sigma, \sigma' \in \Sigma(E^u, E^s; 0)[1]$, abbreviate

$$F = A_{uu} + \phi_u(I_u + \sigma) : E^u \to E^u,$$

$$F' = A_{uu} + \phi_u(I_u + \sigma') : E^u \to E^u;$$

that is,

$$T(\sigma)F = A_{ss}\sigma + \phi_s(I_u + \sigma),$$
$$T(\sigma')F' = A_{ss}\sigma' + \phi_s(I_u + \sigma').$$

Since $F : E^u \to E^u$ is a homeomorphism that fixes the origin, when v runs through $E^u - \{0\}$,

$$\begin{aligned} |T(\sigma) - T(\sigma')|_* &= \sup_{v\neq 0} \frac{|(T\sigma)(v) - (T\sigma')(v)|}{|v|} \\ &= \sup_{v\neq 0} \frac{|(T\sigma)(Fv) - (T\sigma')(Fv)|}{|Fv|}. \end{aligned}$$

On one hand,

$$\begin{aligned} &|(T\sigma)(Fv) - (T\sigma')(Fv)| \\ &\leq |(T\sigma)(Fv) - (T\sigma')(F'v)| + |(T\sigma')(F'v) - (T\sigma')(Fv)| \\ &\leq |A_{ss}(\sigma(v) - \sigma'(v))| + |\phi_s(v + \sigma(v)) - \phi_s(v + \sigma'(v))| + \mathrm{Lip}(T\sigma')|F'v - Fv| \\ &\leq \tau|\sigma(v) - \sigma'(v)| + \mathrm{Lip}\,\phi|\sigma(v) - \sigma'(v)| + \mathrm{Lip}\,\phi|\sigma(v) - \sigma'(v)| \\ &\leq (\tau + 2\mathrm{Lip}\,\phi)|\sigma(v) - \sigma'(v)|. \end{aligned}$$

On the other hand, since $\phi(0) = 0$ and $\sigma(0) = 0$,

$$\begin{aligned} |Fv| &= |A_{uu}v + \phi_u(v + \sigma(v)) - \phi_u(0 + \sigma(0))| \\ &\geq \tau^{-1}|v| - \mathrm{Lip}\,\phi(|v| + \mathrm{Lip}\,\sigma|v|) \\ &\geq (\tau^{-1} - 2\mathrm{Lip}\,\phi)|v|. \end{aligned}$$

Thus

$$\begin{aligned} |T(\sigma) - T(\sigma')|_* &\leq \frac{\tau + 2\mathrm{Lip}\,\phi}{\tau^{-1} - 2\mathrm{Lip}\,\phi} \sup_{v\neq 0} \frac{|\sigma(v) - \sigma'(v)|}{|v|} \\ &= \frac{\tau + 2\mathrm{Lip}\,\phi}{\tau^{-1} - 2\mathrm{Lip}\,\phi} |\sigma - \sigma'|_*. \end{aligned}$$

Since

$$\frac{\tau + 2\mathrm{Lip}\,\phi}{\tau^{-1} - 2\mathrm{Lip}\,\phi} < 1,$$

$T = T_\phi$ is a contraction with respect to the norm $|\cdot|_*$. By the contraction mapping principle, T has a unique fixed point $\sigma = \sigma_\phi \in \Sigma(E^u, E^s; 0)[1]$ such that

$$(A + \phi)(\mathrm{gr}(\sigma)) \subset \mathrm{gr}(\sigma).$$

Note that, for any $(v, \sigma(v)) \in \mathrm{gr}(\sigma)$, letting $u = (A_{uu} + \phi_u(I_u + \sigma))^{-1} v$ gives $(A + \phi)(u, \sigma(u)) = (v, \sigma(v))$. Hence we actually have

$$(A + \phi)(\mathrm{gr}(\sigma)) = \mathrm{gr}(\sigma).$$

Now we prove

$$\mathrm{gr}(\sigma) = W^u(0, A + \phi).$$

We apply Lemma 2.14 to $(A + \phi)^{-1}$ for the case $E(r) = E$ (see the remark after Lemma 2.14). In fact, since $\mathrm{Lip}\,\phi < \delta \le m(A)$, $A + \phi$ is invertible. Moreover, writing $(A + \phi)^{-1} = A^{-1} + \psi$, it is easy to see that if $\mathrm{Lip}\,\phi$ is small, then $\mathrm{Lip}\,\psi$ is small. Hence, shrinking δ if necessary, we may assume $(A + \phi)^{-1}$ satisfies Lemma 2.14. Now since $\sigma(0) = 0$ and $\mathrm{Lip}\,\sigma \le 1$, the graph $\mathrm{gr}(\sigma)$ is contained in the cone $C_1(E^u)$. Since $\mathrm{gr}(\sigma)$ is invariant under $(A + \phi)^{-1}$, by Lemma 2.14,

$$\mathrm{gr}(\sigma) \subset W^u(0, A + \phi).$$

To prove the opposite inclusion, suppose there is $v \in W^u(0, A + \phi) - \mathrm{gr}(\sigma)$. Then there is $w \in \mathrm{gr}(\sigma)$ such that $v_u = w_u$. Then $v - w \notin C_1(E^u)$. By Claim 2 in the proof of Lemma 2.14,

$$|(A + \phi)^{-n}(v) - (A + \phi)^{-n}(w)| \to \infty.$$

But v and w both belong to $W^u(0, A + \phi)$; hence

$$|(A + \phi)^{-n}(v) - (A + \phi)^{-n}(w)| \to 0,$$

a contradiction. This proves $\mathrm{gr}(\sigma) = W^u(0, A + \phi)$, proving item (1).

Item (2) concerns differentiability, which is delicate. To prove it, one way is to use the concept of *hyperbolic set*. This beautiful idea seems due to Pugh. The hyperbolic set is the main concept of Chapter 4, which deserves a full but not a quick preparation. We hence postpone the proof for item (2) to Section 4.3 and stop the proof of Lemma 2.17 here. □

Theorem 2.18 (Local stable manifold for a hyperbolic fixed point). *Let $f : U \to E$ be C^k, $k \ge 1$, and let $0 \in U$ be a hyperbolic fixed point of f with splitting $E = E^s \oplus E^u$. Then there is $r > 0$ such that $W^s_r(0, f)$ is a C^k embedded submanifold of U of dimension $\dim E^s$, tangent at 0 to E^s. Likewise for W^u_r.*

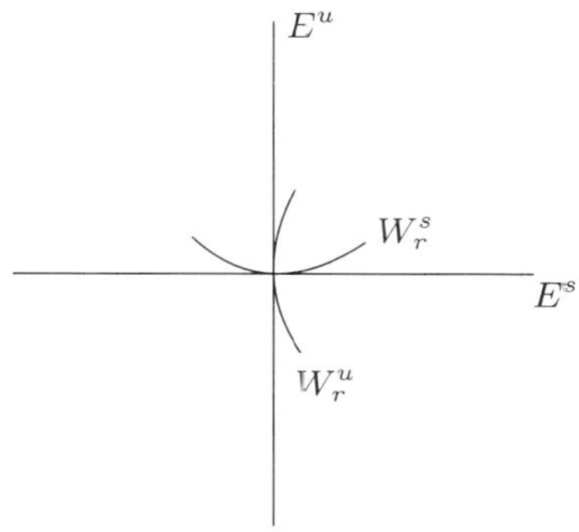

Figure 2.11. The local stable and unstable manifolds.

The proof of Theorem 2.18 is postponed to Section 4.3, page 91.

Remark. Theorems 2.18 and 2.15 are often combined together and referred to as the local stable manifold theorem.

At the end of this chapter we prepare for the need of Section 4.3 an important criterion taken from Katok-Hasselblatt (1995) for testing when a Lipschitz map is differentiable. It uses the notions of secant and tangent lines only.

Let $E = E^u \oplus E^s$, and let $\sigma : E^u \to E^s$ be a Lipschitz map. Denote $G = \mathrm{gr}(\sigma)$. Let $v \in E^u$ be given. Denote $z = (v, \sigma(v))$. Every $h \in E^u$ with $h \neq 0$ determines a *secant line of* G *through* z and $(v + h, \sigma(v + h))$. A line $l \subset T_zE$ through z is called a *generalized tangent line of* G if there is a sequence $h_n \in E^u$ with $h_n \neq 0$ and $h_n \to 0$ such that the sequence of secant lines l_n determined by z and $z_n = (v + h_n, \sigma(v + h_n))$ converges to l. The derivative $Df(z)$ of a diffeomorphism f at z maps a generalized tangent line of G at z to a generalized tangent line of fG at fz. We call the union of all generalized tangent lines of G through z the *tangent set* of G at z, denoted by T_zG. (For instance, for the function $\sigma : \mathbb{R} \to \mathbb{R}$, $\sigma(x) = x \sin(1/x)$, $T_{(0,0)}\mathrm{gr}(\sigma)$ is the union of all lines in $\mathbb{R}^2$ through the origin with slope ≤ 1 in absolute value.)

For every unit vector $e \in E^u$, taking h_n appropriately along the direction of e gives a generalized tangent line l. Since σ is Lipschitz, l is not parallel to E^s. Then the projection $\pi_u : T_zE \to E^u$ maps l onto the line spanned by e. This means $\pi_u(T_zG) \supset E^u$.

Criterion. *Let* $E = E^u \oplus E^s$. *A Lipschitz map* $\sigma : E^u \to E^s$ *is differentiable at* $v \in E^u$ *if and only if the tangent set* T_zG *is contained in a linear subspace of* T_zE *of dimension* $\dim E^u$; *here* $z = (v, \sigma(v))$, $G = \mathrm{gr}(\sigma)$. *Moreover, the tangent set* T_zG *in this case is just the tangent plane of* G *at* z.

Proof. The proof is elementary. It is to interpret differentiability with the geometrical language of secant and generalized tangent lines.

The "only if" part is straightforward. We prove the "if" part. Assume there is a linear subspace V of T_zE of dimension $\dim E^u$ such that $V \supset T_zG$. Let $\pi_u : E \to E^u$ be the projection. As noticed above, $\pi_u(T_zG) \supset E^u$. Then $\pi_u(V) \supset E^u$. Since V and E^u are linear spaces of the same dimension, being a linear map from V onto E^u, $\pi_u|_V$ must be a linear isomorphism. Since $\pi_u|_V$ maps the subset T_zG of V onto E^u, it follows that $T_zG = V$. In particular, for every $e \in E^u$, $e \neq 0$, there is a unique generalized tangent line whose image under π_u contains e. We simply say there is a unique generalized tangent line "above" e. Let

$$L : E^u \to E^s$$

be the unique (linear) map such that $\mathrm{gr}(L) = T_zG$. That is, let

$$L = \pi_s(\pi_u|_{T_zG})^{-1}.$$

Then the generalized tangent line above $e \in E^u$ is just $z + t(e, Le)$.

We prove that σ is differentiable at v, and in fact $D\sigma(v) = L$. We need to prove

$$\lim_{h\to 0} \frac{\sigma(v+h) - \sigma(v) - Lh}{|h|} = 0.$$

Suppose there are $0 \neq h_n \to 0$ such that

$$(*) \qquad \lim_{n\to\infty} \frac{\sigma(v+h_n) - \sigma(v) - Lh_n}{|h_n|} \neq 0.$$

One may assume

$$\lim_{n\to\infty} \frac{h_n}{|h_n|} = e \neq 0.$$

Since σ is Lipschitz, one may also assume

$$\lim_{n\to\infty} \frac{\sigma(v+h_n) - \sigma(v)}{|h_n|} = u.$$

This means that the secant lines through z and $z_n = (v + h_n, \sigma(v + h_n))$ converge to the generalized tangent line $z+t(e, u)$. However, the generalized tangent line above e is $z+t(e, Le)$ and, by $(*)$, $u \neq Le$, a contradiction. This proves the criterion. $\square$

Exercises

Denote E a finite-dimensional normed vector space.

Exercise 2.1. Let $A : E \to E$ be a linear isomorphism. Prove $m(A) = |A^{-1}|^{-1}$.

Exercise 2.2. Prove that all norms on E are equivalent.

Exercise 2.3. Let $A : E \to E$ be a linear isomorphism. Prove that A is hyperbolic if and only if A has no eigenvalue of absolute value 1. (Note that this condition does not involve the norm of E, while the definition of hyperbolicity does. Explain why.)

We give some definitions for the next exercise. Let $A : E \to E$ be a linear isomorphism. Denote

$$B^s = \{v \in E \mid \{|A^n(v)|\}_{n=0}^{\infty} \text{ is bounded}\},$$
$$B^u = \{v \in E \mid \{|A^{-n}(v)|\}_{n=0}^{\infty} \text{ is bounded}\}$$

and

$$D^s = \{v \in E \mid |A^n(v)| \to 0,\ n \to \infty\},$$
$$D^u = \{v \in E \mid |A^{-n}(v)| \to 0,\ n \to \infty\}.$$

Clearly, B^s, B^u, D^s, and D^u are linear subspaces of E, invariant under A.

Exercise 2.4. Let $A : E \to E$ be a linear isomorphism. Prove that the following three conditions are equivalent:

(1) A is hyperbolic;
(2) $B^s \cap B^u = \{0\}$;
(3) $D^s + D^u = E$.

Exercise 2.5. Let $A : E \to E$ be a hyperbolic linear isomorphism. Prove $0 \in E$ is the unique fixed point of A.

Exercise 2.6. Let $A : E \to E$ be a linear map. Denote by $\rho(A)$ the spectral radius of A, that is, the maximal absolute value of eigenvalues of A. Prove that for any $\epsilon > 0$, there is a norm $|\cdot|$ of E such that $|A| < \rho(A) + \epsilon$, where $|A|$ is the operator norm of A with respect to $|\cdot|$.

Exercise 2.7. Show that the set of hyperbolic linear isomorphisms of E forms an open and dense subset of the space of linear isomorphisms of E.

Exercise 2.8. Let $H(\mathbb{R}^n)$ be the set of hyperbolic linear isomorphisms of $\mathbb{R}^n$.

(a) Assume that $A \in H(\mathbb{R}^n)$. Prove there is a curve $\{A_t \in H(\mathbb{R}^n) : 0 \le t \le 1\}$ such that (i) $A_0 = A$ and (ii) A_1 is a real diagonal matrix.

(b) Assume $A \in H(\mathbb{R}^n)$. Prove there is a curve $\{A_t \in H(\mathbb{R}^n) : 0 \le t \le 1\}$ such that (i) $A_0 = A$ and (ii)

$$A_1|_{E^s} = \begin{cases} \operatorname{diag}(1/2, \dots, 1/2) & \text{if } \det(A|_{E^s}) > 0, \\ \operatorname{diag}(1/2, \dots, 1/2, -1/2) & \text{if } \det(A|_{E^s}) < 0 \end{cases}$$

and

$$A_1|_{E^u} = \begin{cases} \operatorname{diag}(2, \dots, 2) & \text{if } \det(A|_{E^u}) > 0, \\ \operatorname{diag}(2, \dots, 2, -2) & \text{if } \det(A|_{E^u}) < 0. \end{cases}$$

Exercise 2.9. Let $B : E \to E$ be a linear isomorphism, represented under a direct sum

$$E = E_1 \oplus E_2$$

as

$$\begin{pmatrix} B_{11} & 0 \\ B_{21} & B_{22} \end{pmatrix}$$

such that

$$\max\{|B_{11}^{-1}|, |B_{22}|\} < 1.$$

Prove B is hyperbolic.

Exercise 2.10. Prove Theorem 2.11.

Exercise 2.11. Let $A_\alpha : \mathbb{R} \to \mathbb{R}$ denote the linear map

$$A_\alpha(x) = \alpha x.$$

(a) Prove that if $0 < \alpha < 1$ and $0 < \beta < 1$, then A_α and A_β are topologically conjugate.

(b) Prove that if $\alpha \neq \beta$, then there is no lipeomorphism $h : \mathbb{R} \to \mathbb{R}$ such that $hA_\alpha = A_\beta h$.

(Thus the topological congugacy h guaranteed by the Hartman-Grobman theorem is generally not a lipeomorphism.)

Exercise 2.12. Let $p \in E$ be a hyperbolic fixed point of f. Given any positive integer m, prove there is a neighborhood V of p such that any period point of f in $V - \{p\}$ has period greater than m.

Chapter 3

Horseshoes, toral automorphisms, and solenoids

In this chapter we introduce the Smale horseshoe, Anosov toral automorphism, and the solenoid attractor. These historic discoveries led to the modern theory of differentiable dynamical systems. We first give some elements of symbolic dynamics, which provide a computable model for the horseshoe.

3.1. Symbolic dynamics

Symbolic dynamics constitutes a deep branch of dynamical systems. Here we just introduce some basic definitions and properties of it.

The *symbolic space* Σ_2 of two symbols is defined to be the bi-infinite product

$$\Sigma_2 = \prod_{n=-\infty}^{\infty} A_n,$$

endowed with the product topology, where $A_n = \{0, 1\}$ for every $n \in \mathbb{Z}$. Thus a point $a \in \Sigma_2$ is a bi-sequence

$$\cdots\ a_{-2}\ a_{-1}\ a_0\ a_1\ a_2\ \cdots$$

where $a_n \in \{0, 1\}$. A basis of neighborhoods of $a \in \Sigma_2$ is formed by sets of the form

$$C_j(a) = \{b \in \Sigma_2 \mid b_n = a_n,\ \forall\ -j \leq n \leq j\},$$

$j = 0, 1, 2, \ldots$. Thus two points $a, b \in \Sigma_2$ are close to each other if they agree on a long interval $[-j, j]$ of center 0. This topology is metrizable by the metric

$$\mathrm{d}(a, b) = \sum_{n=-\infty}^{\infty} \frac{|a_n - b_n|}{2^{|n|}}.$$

Recall that a *Cantor set* is a set that is compact, perfect, totally disconnected, and metrizable. See the paragraph right before Theorem 1.12. Also recall that the *Tychonoff theorem* says that, with the product topology, any product of compact spaces is compact. See for instance Munkres (2000).

Theorem 3.1. *Σ_2 is a Cantor set.*

Proof. By the Tychonoff theorem Σ_2 is compact. Given any $a \in \Sigma_2$ and $j \geq 1$, there is $b \in C_j(a)$ that is different from a. Hence Σ_2 is perfect. We prove Σ_2 is totally disconnected. Take any $a \neq b$ in Σ_2. There is $m \in \mathbb{Z}$ such that $a_m \neq b_m$. Let

$$V(a) = \{c \in \Sigma_2 \mid c_m = a_m\}.$$

Then $V(a)$ is open. Likewise, the complement $\Sigma_2 - V(a)$ is open. Since $a \in V(a)$ and $b \in \Sigma_2 - V(a)$, a and b cannot be in the same connected components. Thus Σ_2 is totally disconnected. □

Define the *shift map* to be

$$\sigma : \Sigma_2 \to \Sigma_2$$
$$(\sigma(a))_n = a_{n+1}, \quad \forall\, n \in \mathbb{Z}.$$

Thus σ shifts every bi-sequence one unit to the left. Clearly σ is a homeomorphism. We call (Σ_2, σ) a *symbolic dynamical system.*

Note that a fixed point of σ is a bi-sequence of constant entries and a periodic point of period k is a bi-sequence with repeated k-tuples. Thus σ has two fixed points, $\cdots 000 \cdots$ and $\cdots 111 \cdots$. Likewise, σ^2 has four fixed points, $\cdots 000 \cdots$, $\cdots 111 \cdots$, $\cdots 01\dot{0}101 \cdots$ and $\cdots 010\dot{1}01 \cdots$, where the dot sign above a digit denotes the 0-place of the bi-sequence. It is easy to calculate the number of periodic points of σ of a given period (Exercise 3.2).

Theorem 3.2. *Periodic points of σ are dense in Σ_2, and σ is transitive on Σ_2.*

Proof. Let $a \in \Sigma_2$ be given. For any $j \geq 1$, let $b \in \Sigma_2$ be the bi-sequence that infinitely repeats the $(2j+1)$-tuple $a_{-j} \cdots a_j$ of a in both directions. Then b is periodic and $b \in C_j(a)$. Thus periodic points are dense.

To show σ is transitive on Σ_2, it suffices to construct a bi-sequence $c \in \Sigma_2$ whose positive orbit is dense in Σ_2. Let $c_n = 0$ for all $n \leq -1$. Starting with $n = 0$, put in successively all possible finite tuples of $0, 1$. More precisely, starting with $n = 0$, first put in the two 1-tuples, 0 and 1. Then put in consecutively the four 2-tuples, $00, 01, 10$, and 11, then the eight 3-tuples, and so on. This defines a point $c \in \Sigma_2$ whose positive part of the bi-sequence contains all finite tuples.

We check that $\mathrm{Orb}^+(c)$ is dense in Σ_2. Take any $a \in \Sigma_2$ and any $j \in \mathbb{N}$. It suffices to prove there is $n \in \mathbb{N}$ such that $\sigma^n(c) \in C_j(a)$. But this is obvious because the $(2j+1)$-tuple $a_{-j} \cdots a_j$ must be somewhere in the positive part of the bi-sequence of c; hence some positive iterate by σ will shift this tuple to the middle. □

The two properties stated in Theorem 3.2 contrast each other. Every periodic orbit is an individual (compact) subsystem. Thus a set with a dense subset of periodic points seems to have a fairly loose structure. On the other hand, a transitive set seems to be tight as a whole. The next theorem indicates that, in case the two properties hold simultaneously, the dynamics of the set must be fairly nontrivial. It holds not only for symbolic dynamics but also for general topological dynamics.

Let $f : X \to X$ be a homeomorphism of a compact metric space. We say f has *sensitive dependence on initial conditions* if there is $r > 0$ such that for any $x \in X$ and any $\delta > 0$, there are $y \in B(x, \delta)$ and $m \geq 1$ such that $d(f^m x, f^m y) \geq r$. Briefly, this means that, in a uniform scale r, at every point $x \in X$, f is not positively Lyapunov stable. (Recall that a point x is *positively Lyapunov stable* for f if for any $r > 0$, there is $\delta > 0$ such that for any $y \in B(x, \delta)$ and any $m \geq 1$, $d(f^m x, f^m y) < r$.) The following theorem is due to Banks et. al. (1992).

Theorem 3.3. *Let X be a compact metric space, and let $f : X \to X$ be a homeomorphism. Assume f has periodic points dense in X and is also transitive on X. If X does not reduce to a single periodic orbit, then f has sensitive dependence on initial conditions.*

Proof. Since periodic points are dense in X and X does not reduce to a single periodic orbit, there are at least two periodic points $p, q \in X$ such that
$$a = d(\mathrm{Orb}(p), \mathrm{Orb}(q)) > 0.$$
We prove that f has sensitive dependence on initial conditions and that the constant r could be $a/8$.

Let $x \in X$ and $\delta > 0$ be given. We may assume $\delta < a/8$. Since x is $a/2$ away from either $\mathrm{Orb}(p)$ or $\mathrm{Orb}(q)$, to be precise we assume
$$d(x, \mathrm{Orb}(p)) \geq a/2.$$

Since periodic points are dense, there is a periodic point $y \in B(x, \delta)$. See Figure 3.1. Assume y has period $k \geq 1$. Let $\eta > 0$ be small enough such

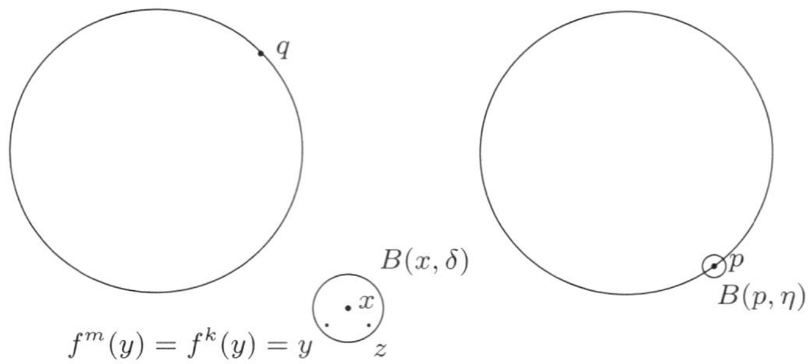

Figure 3.1. Proof of Theorem 3.3. y is periodic of period k; hence it comes back in every k consecutive iterates. On the other hand, z has a dense orbit; hence it can get close to and hence remain close to $\mathrm{Orb}(p)$ for more than k consecutive iterates. Then at some moment they get apart.

that for any $0 \leq i \leq k$,

$$f^i(B(p, \eta)) \subset B(\mathrm{Orb}(p), a/8).$$

Since f is transitive, there are $z \in B(x, \delta)$ and $n \geq 1$ such that $f^n z \in B(p, \eta)$. Then

$$\{f^n z, f^{n+1} z, \ldots, f^{n+k} z\} \subset B(\mathrm{Orb}(p),\ a/8).$$

But some iterate in $\{f^n y, f^{n+1} y, \ldots, f^{n+k} y\}$, say $f^m y$, must be y itself. Hence

$$d(f^m y, f^m z) \geq a/4.$$

Then either $d(f^m y, f^m x) \geq a/8$ or $d(f^m z, f^m x) \geq a/8$. This proves Theorem 3.3. □

3.2. Smale horseshoe

We gave the story of the Smale horseshoe map at the beginning of Chapter 1. Now we study it closely.

Let $Q \subset \mathbb{R}^2$ be a square of size 1. Define a diffeomorphism f so that Q gets contracted horizontally and expanded vertically and folded into a horseshoe-shape and put back across itself. See Figure 3.2.

Since fQ goes beyond Q, some points of Q do not have the second iterates. We hence extend this to a global diffeomorphism $f : S^2 \to S^2$ such that the south pole of the lower hemisphere becomes a source and the upper hemisphere gets mapped into itself. See Figure 3.3.

We focus on Q and single out two horizontal strips H_0 and H_1 and two vertical strips V_0 and V_1 such that

$$fH_0 = V_0,\ \ fH_1 = V_1;$$

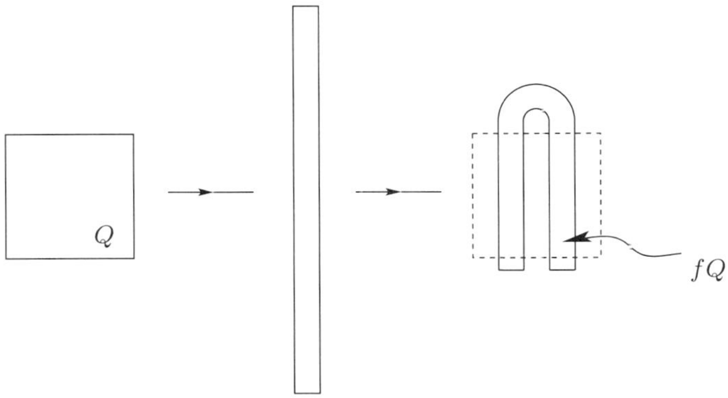

Figure 3.2. The horseshoe map.

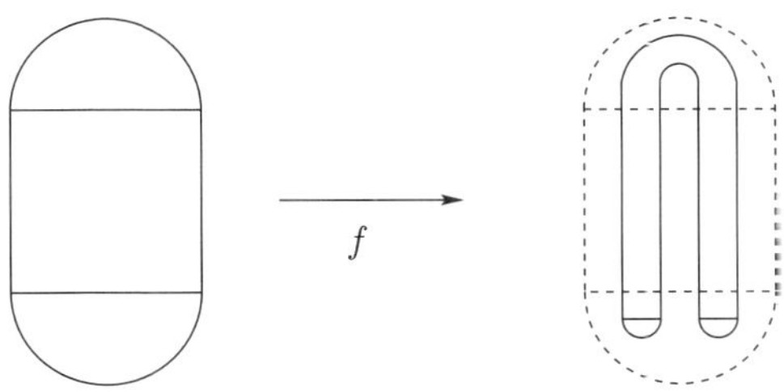

Figure 3.3. The (global) horseshoe map.

see Figure 3.4. For simplicity assume f is affine on H_i with contraction rate 1/5 and expansion rate 5, respectively. We will call a rectangle that crosses H_0 or H_1 horizontally a *horizontal strip* and a rectangle that crosses V_0 or V_1 vertically a *vertical strip*. (Here the terms like "cross" are temporarily used in this proof only, and the meaning will be clear from the contents.)

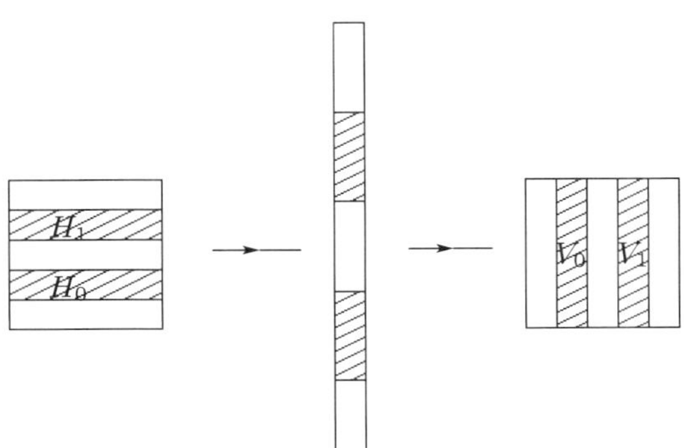

Figure 3.4. H_0, H_1 and V_0, V_1.

Claim 1. *For any vertical strip V, $(fV) \cap V_i$ is a vertical strip with width contracted by $1/5$. For any horizontal strip H, $(f^{-1}H) \cap H_i$ is a horizontal strip with height contracted by $1/5$.*

In fact, the case of a vertical strip is obvious. For a horizontal strip H,

$$(f^{-1}H) \cap H_i = f^{-1}(H \cap fH_i) = f^{-1}(H \cap V_i).$$

Since $H \cap V_i$ is a rectangle that crosses V_i horizontally, its f^{-1}-image is just a horizontal strip. This proves Claim 1.

Now let

$$\Lambda = \bigcap_{n=-\infty}^{\infty} f^n Q.$$

In other words, Λ is the largest f-invariant set contained in Q.

Theorem 3.4 (Smale (1965)). *$f : \Lambda \to \Lambda$ is topologically conjugate to $\sigma : \Sigma_2 \to \Sigma_2$.*

Proof. If $x \notin H_0 \cup H_1$, then $fx \notin Q$. Hence

$$\Lambda = \bigcap_{n=-\infty}^{\infty} f^n(H_0 \cup H_1).$$

Since $H_0 \cap H_1 = \emptyset$, for any $x \in \Lambda$, there is a unique $a \in \Sigma_2$ such that

$$f^n x \in H_{a_n}$$

for all $n \in \mathbb{Z}$. We call this $a \in \Sigma_2$ the *itinerary sequence* of x. Define a map

$$h : \Lambda \to \Sigma_2$$
$$h(x) = \text{the itinerary sequence of } x.$$

Since the itinerary sequence of $f(x)$ is the shift of the itinerary sequence of x, we immediately get

$$hf = \sigma h.$$

We prove h is 1-1 and onto; that is, we prove that for any $a \in \Sigma_2$, there is a unique $x \in \Lambda$ such that $f^n x \in H_{a_n}$ for all $n \in \mathbb{Z}$. We need to prove that, for any $a \in \Sigma_2$,

$$\bigcap_{n=-\infty}^{\infty} f^{-n}(H_{a_n})$$

is a single point. We write this bi-infinite intersection as

$$\cdots \cap f^3 H_{a_{-3}} \cap f^2 H_{a_{-2}} \cap f H_{a_{-1}} \cap H_{a_0} \cap f^{-1} H_{a_1} \cap f^{-2} H_{a_2} \cap \cdots,$$

or

$$\cdots \cap \underbrace{f^2 V_{a_{-3}} \cap \underbrace{f V_{a_{-2}} \cap \underbrace{V_{a_{-1}}}_{J_0}}_{J_1}}_{J_2} \cap \underbrace{\underbrace{\underbrace{H_{a_0}}_{I_0} \cap f^{-1} H_{a_1}}_{I_1} \cap f^{-2} H_{a_2}}_{I_2} \cap \cdots .$$

Clearly

$$I_{n+1} \subset I_n, \quad J_{n+1} \subset J_n.$$

Claim 2. *Every I_n is a horizontal strip with height $\leq 1/5^n$, and every J_n is a vertical strip with width $\leq 1/5^n$.*

In fact, since J_0 is a vertical strip, by Claim 1

$$J_1 = f(V_{a_{-2}}) \cap V_{a_{-1}}$$

is a vertical strip. Hence

$$J_2 = f(f(V_{a_{-3}}) \cap V_{a_{-2}}) \cap V_{a_{-1}}$$

is a vertical strip, and so on. The proof for I_n is similar. This proves Claim 2.

Thus

$$\bigcap_{n=-\infty}^{\infty} f^{-n}(H_{a_n})$$

is an intersection of a horizontal interval and a vertical interval, hence a single point, proving h is 1-1 and onto.

It is easy to see h is continuous. In fact, for any $j \geq 1$, if $x, y \in \Lambda$ are sufficiently close, $f^n x$ and $f^n y$ will be within $1/10$ of each other for all $-j \leq n \leq j$. Hence $f^n x$ and $f^n y$ will be either both in H_0 or else both in H_1, for all $-j \leq n \leq j$; hence the two itinerary sequences $h(x)$ and $h(y)$ will agree on $[-j, j]$. Thus h is continuous. Since Λ is compact, h is a homeomorphism. This proves Theorem 3.4. □

Corollary 3.5. *The horseshoe set Λ is a Cantor set. The horseshoe map $f : \Lambda \to \Lambda$ has periodic points dense and is transitive.*

We briefly indicate that the horseshoe map is structurally stable. Indeed, the geometric construction that leads to the conjugacy h is very coarse: stretching and compressing the square Q, bending it over to cross Q itself. It is intuitively convincing that any diffeomorphism g that is C^1 near f should have the same behavior, hence have a compact invariant set Λ_g that is conjugate to the 2-shift. Therefore $g|_{\Lambda_g}$ is conjugate to $f|_{\Lambda}$.

Let $p \in M$ be a fixed point of f. Define the (global) *stable manifold* and *unstable manifold* of p to be, respectively,

$$W^s(p) = W^s(p, f) = \left\{ x \in M \mid \lim_{n\to+\infty} f^n x = p \right\},$$

$$W^u(p) = W^u(p, f) = \left\{ x \in M \mid \lim_{n\to+\infty} f^{-n} x = p \right\}.$$

It is easy to see that, for any $r > 0$,

$$W^s(p) = \bigcup_{n\geq 0} f^{-n} W_r^s(p),$$

$$W^u(p) = \bigcup_{n\geq 0} f^{n} W_r^u(p).$$

If p is hyperbolic, by Theorem 2.18, the local stable manifold $W_r^s(p)$ (definition on page 45) is a C^1 embedded disc. Thus, as a monotone union of embedded submanifolds, $W^s(p)$ is an immersed submanifold of M.

We say $x \in M$ is a *homoclinic point* of p if

$$x \in W^s(p) \cap W^u(p) - \{p\}.$$

Note that if x is a homoclinic point of p, so will be any point in the orbit of x. A homoclinic point x of p is called *transverse* if

$$T_x W^s(p) \oplus T_x W^u(p) = T_x M.$$

These notions are defined similarly for periodic points.

Transverse homoclinic point is a notion of paramount importance in modern dynamical systems, discovered by H. Poincaré. He realized that a homoclinic point complicates the dynamics dramatically. Since iterates of transverse homoclinic points are transverse homoclinic points, the unstable manifold $W^u(p)$ is forced to double back on itself oscillating faster and faster as it does so. See Figure 3.5. The same is true of $W^s(p)$, with respect to iterates of f^{-1}. They cross each other to give very complicated dynamics. Poincaré (in 1890) has a famous comment about it:

> *The complexity of the figure of a transversal homoclinic orbit is striking and I shall not even try to draw it. Nothing is more suitable to give an idea of the complex nature of the three-body problem, and of all problems in dynamics in general.*

Smale (1967) realized that the best way to understand the transverse homoclinic phenomenon is to embed it in the horseshoe model. Let Λ be the horseshoe set, and let p be the fixed point at the lower-left part of Λ. The way $W^u(p)$ is forced to double back on itself is presented step by step through iterates. We start with the vertical line interval W_0 through p

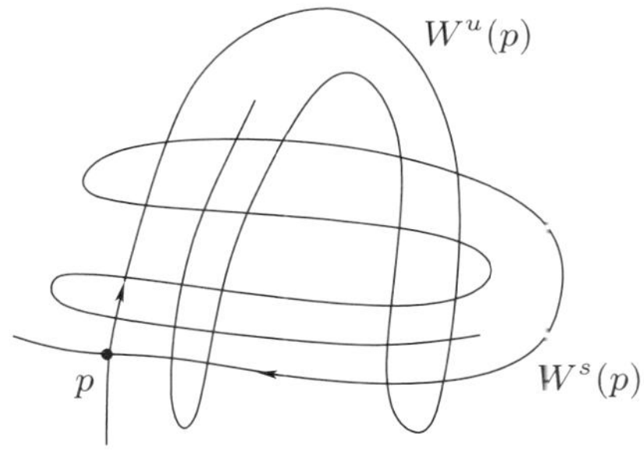

Figure 3.5. Transverse homoclinic phenomenon.

crossing the square Q. See Figure 3.6. It is part of $W^u(p)$. By definition of the horseshoe map, the first iterate of W_0 will be a curve W_1 of shape "U" opening downwards. The second iterate will be more difficult to describe. By definition, we first contract and expand W_1 to a long and thin "U" opening downwards, then bend it to a "double U". This will be W_2.

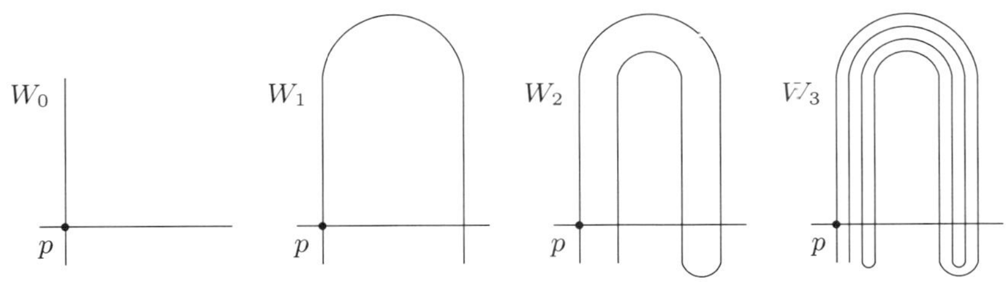

Figure 3.6. The unstable manifold $W^u(p)$ in the horseshoe model. It gets longer and longer under iteration.

To get W_3, we contract and expand W_2 to a long and thin "double U", then bend it to a "double double U", ... It is a valuable experience to complete the picture of W_4. One will see that the picture quickly gets more and more complicated and piled on itself. It is typical for dynamical systems to be complicated but with precise rules.

The following theorem asserts that with a single transverse homoclinic point of f one can recover the whole horseshoe dynamics with respect to some iterate f^m. We state the theorem without proof. See Smale (1965, 1967) or Newhouse (1980).

Theorem 3.6 (Birkhoff-Smale). *Let $p \in M$ be a hyperbolic periodic point of $f : M \to M$, and let x be a transverse homoclinic point of p. For any neighborhood U of $\{p, x\}$, there are $m \geq 1$ and a compact invariant set $\Lambda \subset U$ of f^m such that $\{p, x\} \subset \Lambda$ and such that $f^m : \Lambda \to \Lambda$ is topologically conjugate to $\sigma : \Sigma_2 \to \Sigma_2$.*

3.3. Anosov toral automorphisms

A linear map

$$A : \mathbb{R}^2 \to \mathbb{R}^2$$

is called an *Anosov automorphism* if A is hyperbolic and if A has integer entries and $\det A = \pm 1$. A typical example is

$$\begin{pmatrix} 2 & 1 \\ 1 & 1 \end{pmatrix} : \mathbb{R}^2 \to \mathbb{R}^2.$$

Theorem 3.7. *Let $A : \mathbb{R}^2 \to \mathbb{R}^2$ be an Anosov automorphism. Then the eigenvalues of A are two irrationals with $|\lambda_1| < 1 < |\lambda_2|$, and the slopes of the two eigen-directions are irrational.*

Proof. If the two eigenvalues λ_1 and λ_2 of A are complex conjugate or multiple real roots, they must be of norm 1 as $|\det A| = 1$, contradicting that A is hyperbolic. Thus λ_1 and λ_2 are real and distinct. Since $|\lambda_1\lambda_2| = 1$, we may assume $|\lambda_1| < 1 < |\lambda_2|$.

To see that λ is irrational, suppose by contradiction $\lambda_1 = p/q$, where $p, q \in \mathbb{Z}$ and $(p, q) = 1$. Then

$$p^2 - (a_{11} + a_{22})pq \pm q^2 = 0,$$

where a_{ij} are entries of A. Then $p|q$ and $q|p$. Hence $|\lambda_1| = 1$, contradicting that A is hyperbolic. Thus λ_1 is irrational. Likewise for λ_2.

Let $v \in \mathbb{R}^2$ be an eigenvector of λ_1. Note that $v \neq (0, 1)$ because otherwise $a_{22} = \lambda_1$, contradicting that a_{22} is an integer. Thus v does have slope, and we may assume $v = (1, b)$. Then

$$\begin{pmatrix} a_{11} & a_{12} \\ a_{21} & a_{22} \end{pmatrix} \begin{pmatrix} 1 \\ b \end{pmatrix} = \begin{pmatrix} \lambda_1 \\ \lambda_1 b \end{pmatrix};$$

hence

$$a_{11} + a_{12}b = \lambda_1.$$

Then b is irrational. This proves Theorem 3.7. □

Let $A : \mathbb{R}^2 \to \mathbb{R}^2$ be an Anosov automorphism. Since A has integer entries, it maps $\mathbb{Z}^2$ into itself. Thus $A(a + k) - A(a) = A(k) \in \mathbb{Z}^2$ for any $a \in \mathbb{R}^2$ and any $k \in \mathbb{Z}^2$. Hence A induces a (quotient) map

$$f : \mathbb{T}^2 \to \mathbb{T}^2$$

on the torus such that

$$\pi A = f\pi,$$

where $\pi : \mathbb{R}^2 \to \mathbb{T}^2$ is the projection that takes each component modulo 1. Then π is a local isometry. It is easy to see f is C^∞. Since $\det A = \pm 1$, A^{-1} also has integer entries and hence also induces a C^∞ map of $\mathbb{T}^2$, which is clearly f^{-1}. Thus f is a diffeomorphism of $\mathbb{T}^2$, called an *Anosov toral automorphism.* Anosov made an intensive study of it, which led to the important theory of Anosov diffeomorphisms (named by Smale).

Anosov toral automorphisms are also called *Thom toral automorphisms.* After the horseshoe example was found, Thom gave Smale the toral automorphism as another example of diffeomorphisms with infinitely many periodic orbits that cannot be perturbed away (see Smale (1980) and Shub (1987) for commentary).

The most striking feature for an Anosov toral automorphism f is perhaps the two families of stable and unstable manifolds we now describe.

Let X be a compact metric space, and let $f : X \to X$ be a homeomorphism. Generally, for any point $x \in X$, not necessarily periodic, the (global) *stable manifold* and *unstable manifold* of x with respect to f are defined to be

$$W^s(x, f) = \left\{ y \in X \mid \lim_{n \to +\infty} d(f^n y, f^n x) = 0 \right\},$$

$$W^u(x, f) = \left\{ y \in X \mid \lim_{n \to +\infty} d(f^{-n} y, f^{-n} x) = 0 \right\}.$$

Clearly,

$$f(W^s(x)) = W^s(fx), \; f(W^u(x)) = W^u(fx).$$

Note that if $y \in W^s(x)$, then $x \in W^s(y)$ and $W^s(x) = W^s(y)$. In fact stable manifolds are equivalence classes of the equivalence relation $\sim$, where $x \sim y$ if $d(f^n(y),\; f^n(x)) \to 0$. Hence X decomposes into a disjoint union of stable manifolds. If x is periodic, it reduces to the previous definitions.

The next theorem states that, for Anosov toral automorphisms, $W^s(x)$ and $W^u(x)$ form two beautiful families of immersed C^∞ submanifolds that fill out $\mathbb{T}^2$. See Figure 3.7.

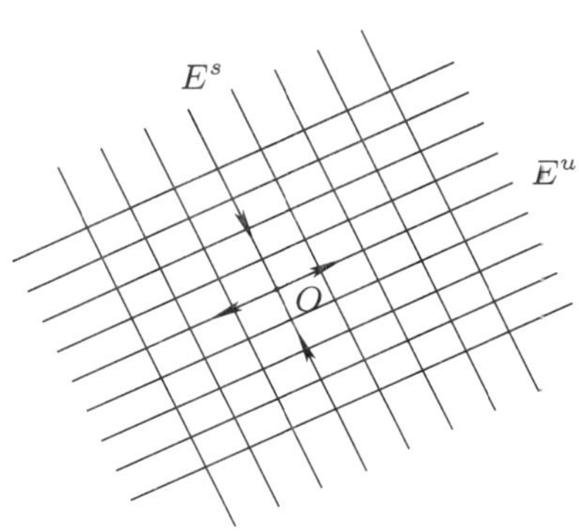

Figure 3.7. The stable and unstable manifolds of A.

Theorem 3.8. *Let $f : \mathbb{T}^2 \to \mathbb{T}^2$ be an Anosov toral automorphism induced by $A : \mathbb{R}^2 \to \mathbb{R}^2$.*

(1) *For any $a \in \mathbb{R}^2$, $W^s(a, A) = a + E^s$, where $\mathbb{R}^2 = E^s \oplus E^u$ is the hyperbolic splitting of A.*

(2) *For any $x \in \mathbb{T}^2$, $W^s(x, f) = \pi(W^s(a, A))$, where a is any point of $\pi^{-1}(x)$.*

(3) *$W^s(x, f)$ is an immersed C^∞ submanifold that is dense in $\mathbb{T}^2$. Likewise for $W^u(x)$. For any $x, y \in \mathbb{T}^2$, $W^s(x, f)$ intersects $W^u(y, f)$ transversely at a dense subset of $\mathbb{T}^2$.*

Proof. (1) Let $b \in W^s(a, A)$. Then $|A^n b - A^n a| \to 0$, $|A^n(b - a)| \to 0$, which means $b - a \in E^s$, or $b \in a + E^s$. This proves $W^s(a, A) \subset a + E^s$. Each step of the argument is invertible; hence the converse is true. This proves $W^s(a, A) = a + E^s$.

(2) Let $x = \pi a$. First we prove $\pi(W^s(a, A)) \subset W^s(x, f)$. Let $b \in W^s(a, A)$. Then $|A^n b - A^n a| \to 0$. Since π is uniformly continuous, we have $d(\pi(A^n b),\ \pi(A^n a)) \to 0$, $d(f^n(\pi b),\ f^n(x)) \to 0$. This means $\pi b \in W^s(x, f)$, proving $\pi(W^s(a, A)) \subset W^s(x, f)$.

Next we prove $\pi(W^s(a, A)) \supset W^s(x, f)$. Take $\epsilon > 0$ (for instance $\epsilon = 1/2$) such that π preserves distance within ϵ; namely, for any $a, b \in \mathbb{R}^2$, if $|a - b| \le \epsilon$, then $d(\pi a,\ \pi b) = |a - b|$. Take $0 < \delta < \epsilon$ such that, for any $a, b \in \mathbb{R}^2$, if $|a - b| \le \delta$, then $|Aa - Ab| \le \epsilon$.

Let $y \in W^s(x, f)$. We prove there is $b \in W^s(a, A)$ such that $\pi b = y$. Take $m \in \mathbb{N}$ sufficiently large such that

$$d(f^n y, f^n x) \le \delta, \quad \forall\, n \ge m.$$

Since $\pi(A^m a) = f^m x$, there is a unique $c \in B(A^m a, \epsilon)$ such that $\pi c = f^m y$. Let $b = A^{-m} c$. See Figure 3.8. Then

$$\pi b = \pi(A^{-m} c) = f^{-m}(\pi c) = f^{-m} f^m y = y.$$

It remains to prove $b \in W^s(a, A)$, that is, to prove $c \in W^s(A^m a, A)$. Since $|c - A^m a| \le \epsilon$, we have

$$|c - A^m a| = d(\pi c,\ \pi(A^m a)) = d(f^m y,\ f^m x) \le \delta.$$

Then

$$|Ac - A(A^m a)| \le \epsilon.$$

Then

$$|Ac - A(A^m a)| = d(\pi(Ac),\ \pi(A(A^m a))) = d(f^{m+1} y,\ f^{m+1} x) \le \delta.$$

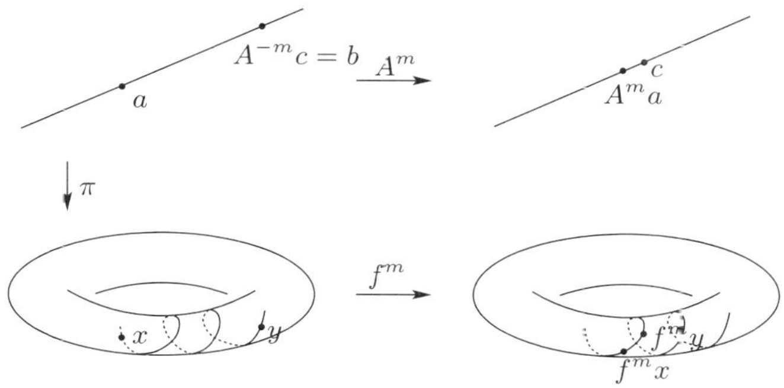

Figure 3.8. The proof of item (2) in Theorem 3.8. Wait long enough so that $f^m y$ gets close to $f^m x$ and hence can be lifted to the right π-preimage c in $\mathbb{R}^2$.

Inductively,

$$|A^n c - A^n(A^m a)| = d(f^{m+n}y,\ f^{m+n}x) \leq \delta.$$

But $d(f^{m+n}y,\ f^{m+n}x) \to 0$; hence $c \in W^s(A^m a, A)$.

(3) Since $W^s(a, A)$ is a line in $\mathbb{R}^2$ and $\pi : \mathbb{R}^2 \to \mathbb{T}^2$ is a C^∞ local embedding, $W^s(x, f)$ is an immersed C^∞ submanifold of $\mathbb{T}^2$. We prove it is dense in $\mathbb{T}^2$. For simplicity we take the case $x = \pi(0)$. Then $W^s(0, A)$ is just E^s, which is by Theorem 3.7 a line through the origin with an irrational slope b. Note that each vertical line through a point of $\mathbb{Z}^2$ represents, say, the latitude circle S^1 of $\mathbb{T}^2$. Now E^s cuts these lines at heights $\{nb\}_{n\in\mathbb{Z}}$ which, modulo integers, is a dense subset of S^1 (Example 3 of Section 1.3). Thus $W^s(x, f)$ is dense in $\mathbb{T}^2$. The rest of the conclusions are obvious. This proves Theorem 3.8. □

Theorem 3.9. *Let $f : \mathbb{T}^2 \to \mathbb{T}^2$ be an Anosov toral automorphism. Then periodic points of f are dense in $\mathbb{T}^2$, and f is transitive on $\mathbb{T}^2$.*

Proof. To prove that periodic points of f are dense in $\mathbb{T}^2$, it suffices to prove that any "rational point" $x \in \mathbb{T}^2$ is f-periodic. Let $a = (p_1/q_1,\ p_2/q_2) \in \mathbb{R}^2$. For any $n \geq 1$, A^n is a matrix with integer entries. Hence the two components of $A^n a$ are both rational with denominator not beyond $[q_1, q_2]$. Then there must be $n > m \geq 0$ such that

$$A^n a - A^m a \in \mathbb{Z}^2.$$

Hence

$$\pi A^n a = \pi A^m a.$$

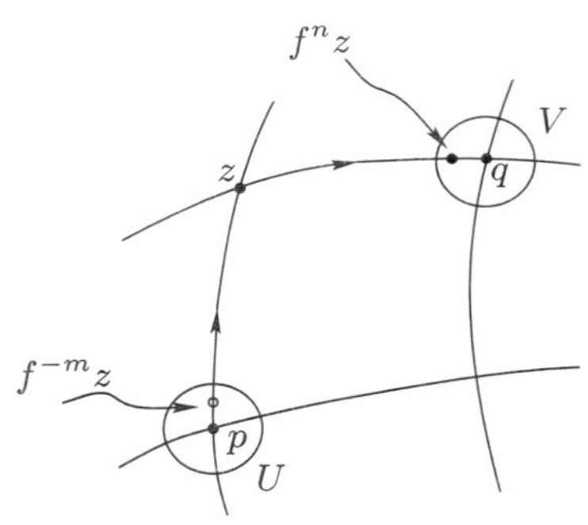

Figure 3.9. The proof of the transitivity of f. The point $f^{-m}z$ goes from U to V.

Denote $x = \pi a$. Then

$$f^n x = f^m x,$$
$$f^{n-m}(x) = x.$$

Hence x is f-periodic. This proves that periodic points of f are dense in $\mathbb{T}^2$.

To prove f is transitive on $\mathbb{T}^2$ we use Theorem 1.6. Let U and V be two open subsets of $\mathbb{T}^2$. Since periodic points are dense in $\mathbb{T}^2$, there are periodic points $p \in U$ and $q \in V$. See Figure 3.9. By Theorem 3.8,

$$W^u(p) \cap W^s(q) \neq \emptyset.$$

Since we are to find a point z that goes from U to V, we may switch to an iterate, or simply assume that p and q are fixed points of f. Let $z \in W^u(p) \cap W^s(q)$. Then there are $m, n \geq 1$ such that $f^{-m}z \in U$ and $f^n z \in V$. Then $(f^{m+n}(U)) \cap V \neq \emptyset$. By Theorem 1.6, f is transitive on $\mathbb{T}^2$. This proves Theorem 3.9. □

Anosov toral automorphisms are structurally stable. In fact, Anosov (1967) proves that *Anosov diffeomorphisms*, among which the "linear ones" defined in this section are special models, are structurally stable. In Chapter 4 we will state and prove this celebrated theorem.

3.4. The solenoid attractor

Recall from Chapter 1 that a compact invariant set A of f is called an attracting set if A has an open neighborhood U with $f(\overline{U}) \subset U$ such that $A = \bigcap_{n\geq 0} f^n(\overline{U})$. An attracting set is called an *attractor* if it is transitive. An attractor that is not a single periodic orbit is called *nontrivial*. In this section we introduce a nontrivial attractor, the solenoid.

The solenoid was attributed to people in topology. Smale (1967) introduced it as an attractor to dynamical systems. Williams (1967, 1974) developed a general theory of attractors.

Let $D^2 = \{z \in \mathbb{C} : |z| \leq 1\}$ be the unit disc in the complex plane. Let $T = S^1 \times D^2 \subset \mathbb{R}^3$ be the solid torus. We stretch T twice as long and compress it 1/4 as thin and then put it back into T, wrapping it twice around. See Figure 3.10.

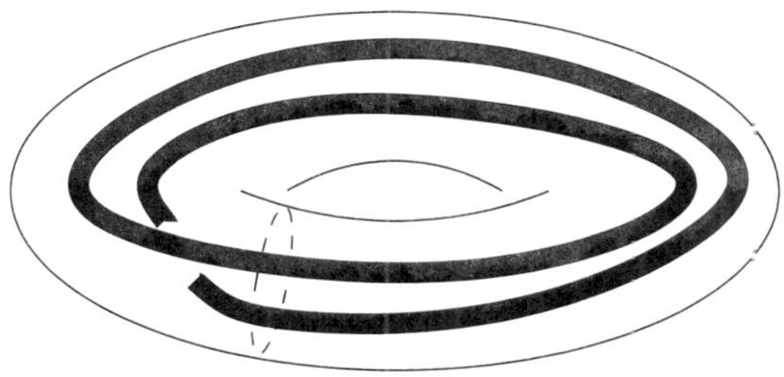

Figure 3.10. The solenoid attractor.

We may write an explicit formula for this map:

$$f : T \to T$$
$$f(t, z) = (g(t),\ z/4 + e^{2\pi t i}/2),$$

where $g : S^1 \to S^1$ is the doubling map $g(t) = 2t \pmod 1$. Let

$$\pi : T \to S^1$$
$$\pi(t, z) = t$$

be the projection. Then

$$\pi f = g\pi.$$

Denote by

$$D(t) = \{t\} \times D^2, \qquad t \in S^1,$$

the slice of T through t. Then f maps $D(t)$ to a disc of radius 1/4 in $D(gt)$, and f^2 maps $D(t)$ to a disc of radius 1/16 in $D(g^2t)$. In general, f^n maps $D(t)$ to a disc of radius $1/4^n$ in $D(g^nt)$. It is clear that $f(T) \subset \operatorname{int}(T)$. The attracting set

$$A = \bigcap_{n \geq 0} f^n(T)$$

is called the *solenoid*.

Note that $f^n(T) \subset f^{n-1}(T)$ is a solid torus that wraps around T for 2^n times. For fixed $t \in S^1$, $f(T)$ crosses $D(t)$ twice, giving two disjoint discs of radius 1/4 in $D(t)$, and $f^2(T)$ crosses $D(t)$ four times, giving four disjoint discs of radius 1/16 in $D(t)$. In general, $f^n(T)$ crosses $D(t)$ 2^n times, giving 2^n disjoint discs of radius $1/4^n$ in $D(t)$. See Figure 3.11. The intersection of a nested sequence of such discs in $D(t)$ gives one point of A in $D(t)$. Thus $A \cap D(t)$ is a Cantor set.

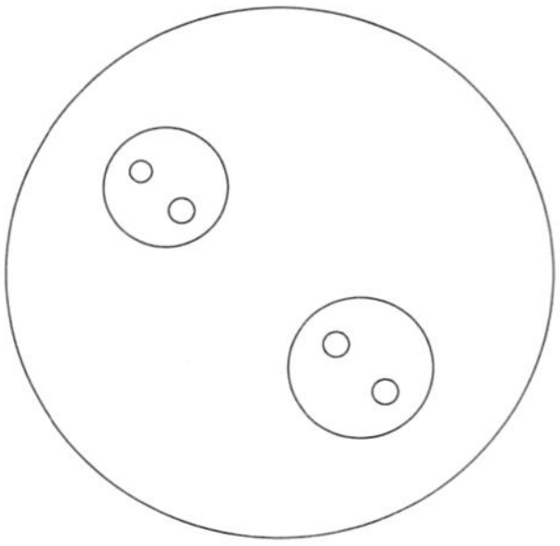

Figure 3.11. Sections of the solenoid.

For fixed $a, b \in S^1$, the sequence of solid tori $\{f^n(T)\}$ is cut by $D(a)$ and $D(b)$ to tubes of length $b - a$. The intersection of any nested sequence of these tubes is an interval of length $b-a$ in A. Thus A is locally a Cantor set times an interval. Globally, A is a union of immersed lines. Every line is not a closed curve because otherwise iterates of f^{-1} would contract some closed curve, taking it to neighborhoods of some points of Λ, contradicting the local structure of the Cantor times an interval.

Explicitly, the tube of T cut by $D(a)$ and $D(b)$ is

$$D[a,b] = \bigcup_{t \in [a,b]} D(t).$$

It is a tube of radius 1 and length $b - a$. Clearly, $f^n(D[a,b])$ is a tube of $f^n(T)$ of radius $1/4^n$ and length $2^n(b-a)$.

Every $x \in A$ has arbitrarily small neighborhoods in T of the form $f^n(D[a,b])$. More precisely,

Fact. *For any $\delta > 0$, there is $n \in \mathbb{N}$ such that, for any $x \in A$, there are $a, b \in S^1$ such that $x \in \operatorname{int} f^n(D[a,b]) \subset B(x,\delta)$.*

In fact, take $n \in \mathbb{N}$ large such that $2^{-n} < \delta$. For $x \in A$, let $t = \pi(f^{-n}(x))$, and let $a = t - 2^{-n}\delta$ and $b = t + 2^{-n}\delta$. It is straightforward to check that this choice works.

Lemma 3.10. *Periodic points of g are dense in S^1.*

Proof. Take any interval $[a,b] \subset S^1$. There is $n \geq 0$ large such that $g^n[a,b]$ covers the whole S^1. Hence there is a subinterval $[a',b'] \subset [a,b]$ such that g^n maps $[a',b']$ homeomorphically onto $[a,b]$, giving a fixed point of g^n in $[a',b']$. □

Theorem 3.11. *The solenoid A has the following properties:*

(1) *Periodic points of f are dense in A.*

(2) *$f|_A$ is transitive.*

Proof. (1) Let $x \in A$. Take any neighborhood U of x in T. We prove there is a periodic point p of f in U. Then p is in A since T is a trapping region.

We may assume $U = f^n(D[a,b])$ for some n large and $b-a$ small. Then it suffices to prove there is a periodic point of f in $D[a,b]$. By Lemma 3.10, there are $t \in [a,b]$ and $k \geq 1$ such that $g^k(t) = t$. Then f^k maps $D(t)$ into itself, giving a fixed point of f^k in $D(t)$.

(2) Let U, V be two open subsets of A. By the Birkhoff Theorem, Theorem 1.6, it suffices to prove there is $k \geq 1$ such that

$$f^k(U) \cap V \neq \emptyset.$$

There are two open subsets U' and V' of T such that

$$U = U' \cap A, \quad V = V' \cap A.$$

Also, there is a choice of $n \in \mathbb{N}$ and $a, b, c, d \in S^1$ such that

$$f^n(D[a,b]) \subset U', \quad f^n(D[c,d]) \subset V'.$$

Hence it suffices to prove there is $k \geq 1$ such that

$$f^k(f^n(D[a,b]) \cap A) \cap (f^n(D[c,d]) \cap A) \neq \emptyset.$$

By applying f^{-n}, it reduces to proving that there is $k \geq 1$ such that

$$f^k(D[a,b]) \cap D[c,d] \cap A \neq \emptyset.$$

But this is obvious because, if k is large, then $g^k([a,b])$ covers S^1, so $f^k(D[a,b])$ crosses $D[c,d]$. Then $f^k(D[a,b]) \cap D[c,d]$ contains a tube of $f^k(T)$ (of length $d-c$), certainly intersecting A. This proves Theorem 3.11. □

Exercises

Exercise 3.1. Let d_* be the following metric on Σ_2:

$$d_*(a,b) = \sum_{n=-\infty}^{\infty} \frac{|a_n - b_n|}{n^2}.$$

Show that it determines the same topology as the metrics $d(a,b)$ in Section 3.1.

Exercise 3.2. Let $\sigma : \Sigma_2 \to \Sigma_2$ be the shift map. Count the number of fixed points of σ^n.

Exercise 3.3. Let $A = (a_{ij})$ be a 2×2 matrix whose entries are either zeroes or ones. Let

$$\Sigma_A = \{s \in \Sigma_2 \mid a_{s_n s_{n+1}} = 1, \ \forall\, n \in \mathbb{Z}\}.$$

Show that Σ_A is σ-invariant. We call the restriction $\sigma_A = \sigma\,|_{\Sigma_A}$ a *subshift of finite type.*

Exercise 3.4. Let A be a $0, 1$ matrix. We call A *eventually positive* if, for some positive integer m, all entries of A^m are positive. Show that if A is eventually positive, then the subshift of finite type σ_A is topologically mixing and its periodic points are dense in Σ_A.

Exercise 3.5. Prove that the shift map $\sigma : \Sigma_2 \to \Sigma_2$ has uncountably many disjoint minimal sets.

Exercise 3.6. Let $f : S^2 \to S^2$ be the global horseshoe map. Find a fixed point p of f, and draw the figure of the unstable manifold $W^u(p)$, as precisely as possible, to see how $W^u(p)$ wraps around. Do the same for the stable manifold $W^s(p)$.

Exercise 3.7. Let Λ be the horseshoe Cantor set of the global horseshoe map f in the previous problem. Prove $\Lambda = \overline{W^u(p)} \cap \overline{W^s(p)}$, where p is any periodic point in Λ.

Exercise 3.8. Prove the following:

(1) The Smale horseshoe is mixing.

(2) The Anosov automorphism $\begin{pmatrix} 2 & 1 \\ 1 & 1 \end{pmatrix}$ is mixing.

Exercise 3.9. Let f_A be an Anosov toral automorphism on $\mathbb{T}^2$ induced by an Anosov automorphism A on R^2. Show that the number of fixed points of f_A equals $|\det(A - I)|$.

Exercise 3.10. Let f be an Anosov toral automorphism on $\mathbb{T}^2$. Let g be a homeomorphism of $\mathbb{T}^2$ which is homotopic to f. Prove there exist an onto continuous map $h : \mathbb{T}^2 \to \mathbb{T}^2$ such that $h \circ g = f \circ h$.

Chapter 4

Hyperbolic sets

Hyperbolic set theory is the analytic foundation for the theory of structural stability. The classical examples of hyperbolic sets are the Smale horseshoe, Anosov toral automorphisms, and the solenoid attractor, introduced in the last chapter. The main feature exhibited in these examples is a uniform contraction in one direction and a uniform expansion in the other direction. Thus every point in the set is like a saddle, but moving, perhaps nonperiodically. While the appearance of finitely many periodic saddles is common, the surprising feature that infinitely many periodic saddles, with unbounded periods but uniform rates of contraction and expansion, fit together harmoniously and even structurally stably in a compact invariant set was not known until the early 1960s. Such a phenomenon can only be "saddle-like". In fact, a uniform contraction in all directions (sink-like) or a uniform expansion in all directions (source-like), on a compact set, must reduce to finitely many periodic orbits (Theorem 4.1).

4.1. The concept of hyperbolic set

Let M be a compact C^∞ Riemannian manifold without boundary, and let $f : M \to M$ be a *diffeomorphism*. By definition this means f is a homeomorphism such that both f and f^{-1} are C^1.

An invariant set $\Lambda \subset M$ of f is called *hyperbolic* if, for each $x \in \Lambda$, the tangent space T_xM splits into a direct sum

$$T_xM = E^s(x) \oplus E^u(x),$$

invariant (as family) in the sense that

$$Tf(E^s(x)) = E^s(f(x)),\ Tf(E^u(x)) = E^u(f(x))$$

such that, for some constants $C \geq 1$ and $0 < \lambda < 1$, the following uniform estimates hold:

$$|Tf^n(v)| \leq C\lambda^n|v|, \ \forall x \in \Lambda, \ v \in E^s(x), \ n \geq 0,$$
$$|Tf^{-n}(v)| \leq C\lambda^n|v|, \ \forall x \in \Lambda, \ v \in E^u(x), \ n \geq 0.$$

In particular, if Λ is a single orbit, it will be called a *hyperbolic orbit.*

The notion of hyperbolic set was introduced by Smale (1967) based on the work of Anosov (1967). It is a fundamental concept for modern dynamical systems and is perhaps the most important concept in this book. We give some remarks.

Remark. (1) Here we have used the notation of the tangent map Tf, with base points x dropped. Compared with the notation of the derivative $Df(x)$ that marks the base point x, they are related as

$$Tf(v) = T_x f(v) = Df(x) \cdot v,$$

where $x = \pi v$. Here π denotes the bundle projection; that is, $x = \pi v$ if and only if $v \in T_x M$. Thus the base point x follows automatically and hence can be dropped from the notation. Note that $T(f^n) = (Tf)^n$; hence there would be no confusion in writing Tf^n. Here $|\cdot|$ denotes the norm (Finsler structure) induced by the Riemannian metric of M. We have also dropped base points from the norm notation. Thus $|v|$ automatically means $|v|_x$, where $x = \pi v$. Likewise, $|Tf^n(v)|$ automatically adopts the norm $|\cdot|_{f^n x}$ at $f^n x$.

(2) Since M is compact, the hyperbolicity of Λ is independent of the choice of the Riemannian metric of M.

(3) The dimension $\dim E^s(x)$ is constant along every orbit of x, called the *index* of the orbit.

(4) $E^s(x)$ or $E^u(x)$ could be $\{0\}$. In this case Λ is said to be of *expanding type* or *contracting type*, respectively.

(5) If Λ is hyperbolic with respect to f, then it is hyperbolic with respect to f^{-1}.

(6) By invariance, the two inequalities in the definition involve not only positive but in fact all iterates of v. That is, for any $x \in \Lambda$ and any $-\infty < m < \infty$,

$$|Tf^n(Tf^m v)| \leq C\lambda^n|Tf^m v|, \ \forall v \in E^s(x), \ n \geq 0.$$

In particular, letting $m = -n$ gives

$$|Tf^{-n}v| \geq C^{-1}(\lambda^{-1})^n|v|, \ \forall v \in E^s(x), \ n \geq 0.$$

Likewise for E^u.

(7) Any invariant subset of a hyperbolic set is hyperbolic. A finite union of hyperbolic sets is hyperbolic (by taking the maximum of the constants

C and λ). An infinite union may break the hyperbolicity. In fact, the key point in the definition of hyperbolic set is the *uniformness* of the constants C, λ.

(8) A hyperbolic fixed point or periodic orbit is a hyperbolic set. The Smale horseshoe, the Anosov toral automorphisms, and the solenoid attractor are more complicated hyperbolic sets. A diffeomorphism f is called an *Anosov diffeomorphism* if the whole manifold M is a hyperbolic set of f. Anosov toral automorphisms are standard examples of Anosov diffeomorphisms. A very enlightening example of a hyperbolic set is the orbit of a transverse homoclinic point. More precisely, let p be a hyperbolic fixed point, and let x be a transverse homoclinic point of p. Then $\mathrm{Orb}(x)$ is a hyperbolic set with invariant splitting obtained by iterating the splitting $T_xM = T_xW^s(p) \oplus T_xW^u(p)$ (Exercise 4.2).

The Riemannian metric on the tangent bundle TM induces a metric on M by defining $d(x, y)$ to be the infimum of the lengths of piecewise differentiable curves joining x and y. As usual, define

$$B(x, r) = \{y \in M \mid d(y, x) \leq r\}.$$

We first prove that a hyperbolic set of contracting type is trivial in the sense that it contains finitely many points. Likewise for the expanding case.

Theorem 4.1. *Let $\Lambda \subset M$ be a hyperbolic set of f of contracting type. Then Λ consists of finitely many periodic orbits of f.*

Proof. Let $C \geq 1$ and $0 < \lambda < 1$ be the hyperbolic constants of Λ. Thus, for every $x \in \Lambda$,

$$|Df^n(x)| \leq C\lambda^n, \quad \forall\, n \geq 1.$$

This inequality holds for every $z \in \overline{\Lambda}$. In fact, let $x_k \in \Lambda$, $x_k \to z$. Then

$$|Df^n(x_k)| \leq C\lambda^n, \quad \forall\, n \geq 1.$$

Fixing n and letting $k \to \infty$ then yields the conclusion. Thus we may assume that Λ is compact.

Fix N and μ such that

$$C\lambda^N < \mu < 1.$$

Let $g = f^N$. Then $g(\Lambda) = \Lambda$ and

$$|Dg(x)| \leq C\lambda^N < \mu$$

for every $x \in \Lambda$. It reduces to proving that Λ consists of finitely many (contracting) periodic orbits of g. For this purpose we must investigate the neighborhoods of every point $x \in \Lambda$.

Claim. *There is $r > 0$ such that, for any $x \in \Lambda$, $\mathrm{Lip}(g) \leq \mu$ on $B(x, r)$.*

In fact, there is $r > 0$ such that, for any $x \in \Lambda$ and any $z \in B(x,r)$, $|Dg(z)| \leq \mu$. The claim then follows from the generalized mean value theorem.

Remark. This implies that a contracting periodic point is isolated in the nonwandering set. In particular, it is isolated in the set of periodic points.

Thus it suffices to prove that every $x \in \Lambda$ is a (contracting) periodic point of g because then, being compact, Λ must reduce to finitely many points and hence must consist of finitely many (contracting) periodic orbits of g.

Take any $x \in \Lambda$. We first prove that every $y \in \alpha(x,g)$ is a contracting periodic point of g. Take $n_i \to \infty$ such that $z_i = g^{-n_i}(x) \to y$. Let $m_i = n_{i+1} - n_i$. By the claim, g^{m_i} is a μ^{m_i}-contracting mapping on $B(z_{i+1}, r)$. Since r is independent of i and $d(z_i, z_{i+1}) \to 0$, there is a large i such that $y \in B(z_{i+1}, r)$ and $B(z_i, \mu^{m_i} r) \subset B(z_{i+1}, r)$. Then g^{m_i} has in $B(z_{i+1}, r)$ a unique fixed point p, which is of contracting type. Since p is the unique nonwandering point of g in $B(z_{i+1}, r)$, it follows that $y = p$. This proves that y is a contracting periodic point of g.

It is easy to see that no point can approach a contracting periodic point under negative iterates, except points of the periodic orbit themselves (see Exercise 4.4). Hence $x \in \mathrm{Orb}(y,g)$. Thus x is a contracting periodic point of g, proving Theorem 4.1. □

All the results of this book on hyperbolic sets hold trivially for the contracting and expanding cases. We will not state the results separately for these two cases.

Let $\Lambda \subset M$ be a hyperbolic set of f. For $x \in \Lambda$, $\gamma > 0$, denote by

$$C_\gamma(E^s(x)) = \{v \in T_x M \mid |v_u| \leq \gamma |v_s|\},$$

$$C_\gamma(E^u(x)) = \{v \in T_x M \mid |v_s| \leq \gamma |v_u|\}$$

the *γ-cones at x about $E^s(x)$* and $E^u(x)$, respectively.

The next theorem corresponds to Theorem 2.2.

Theorem 4.2 (Characterization of E^s). *Let $\Lambda \subset M$ be a hyperbolic set of f with splitting $T_\Lambda M = E^s \oplus E^u$. For any $x \in \Lambda$, $E^s(x)$ is characterized by*

$$\begin{aligned} E^s(x) &= \{v \in T_x M \mid |Tf^n v| \to 0, n \to +\infty\} \\ &= \{v \in T_x M \mid \exists r > 0 \text{ such that } |Tf^n v| \leq r, \forall n \geq 0\} \\ &= \{v \in T_x M \mid \exists \gamma > 0 \text{ such that } Tf^n v \in C_\gamma(E^s(f^n x)), \forall n \geq 0\}. \end{aligned}$$

Likewise,

$$\begin{aligned}E^u(x) &= \{v \in T_xM \mid |Tf^{-n}v| \to 0, n \to +\infty\} \\ &= \{v \in T_xM \mid \exists r > 0 \text{ such that } |Tf^{-n}v| \leq r, \forall n \geq 0\} \\ &= \{v \in T_xM \mid \exists \gamma > 0 \text{ such that } Tf^{-n}v \in C_\gamma(E^u(f^{-n}x)), \forall n \geq 0\}.\end{aligned}$$

In particular, hyperbolic splitting is unique. That is, if $T_xM = G^s(x) \oplus G^u(x)$, $x \in \Lambda$, is another hyperbolic splitting of f on Λ, then $G^s(x) = E^s(x)$, $G^u(x) = E^u(x)$.

Proof. The proof is the same as that of Theorem 2.2, just involving base points of vectors. For instance we prove for $E^s(x)$ that the third set is contained in the fourth. In fact, if

$$v \in T_xM - \{v \in T_xM \mid \exists \gamma > 0 \text{ such that } Tf^n v \in C_\gamma(E^s(f^n x)), \forall n \geq 0\},$$

then there is $m \geq 0$ such that $w = Tf^m v \in T_{f^m x}M - C_1(E^s(f^m x))$. In particular, $w_u \neq 0$. Then

$$|Tf^n(w_u)| \to \infty, \quad |Tf^n(w_s)| \to 0$$

as $n \to +\infty$. Hence

$$|Tf^n w| \geq |Tf^n(w_u)| - |Tf^n(w_s)| \to \infty.$$

Hence $\{|Tf^n v|\}_{n=0}^{+\infty}$ is unbounded, proving that the third set is contained in the fourth. The proofs for the other inclusions are omitted.

The uniqueness of hyperbolic splitting follows immediately because $E^s(x)$ and $E^u(x)$ have been characterized as sets. This proves Theorem 4.2. □

The linear subspace $E^s(x)$ varies continuously with the base point x. To make this precise, we introduce a topology on the set of linear subspaces.

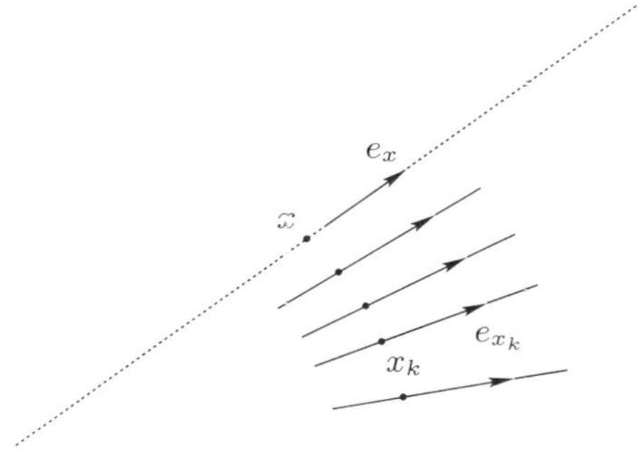

Figure 4.1. Convergence of linear subspaces.

Let $m \geq 1$ be an integer. Let $x \in M$ be a point, and let $E(x)$ be an m-dimensional linear subspace of T_xM. Let $x_k \in M$ be a sequence, and let $E(x_k)$ be an m-dimensional linear subspace of $T_{x_k}M$. We say the $E(x_k)$ *converge to* $E(x)$, written as $E(x_k) \to E(x)$, if there are a basis $\{e^1_{x_k}, \ldots, e^m_{x_k}\}$ of $E(x_k)$ for every k and a basis $\{e^1_x, \ldots, e^m_x\}$ of $E(x)$ such that $e^1_{x_k} \to e^1_x, \ldots, e^m_{x_k} \to e^m_x$. See Figure 4.1. This gives a topology for the m-Grassmann space

$$G^m(M) = \{V \mid V \text{ is an } m\text{-dimensional linear subspace of } T_xM,\ x \in M\}$$

of M. Clearly, $G^m(M)$ is compact. Note that $E(x_k) \to E(x)$ implies $x_k \to x$ and, for any $v \in E(x)$, there are $v_k \in E(x_k)$ such that $v_k \to v$.

Theorem 4.3. *Let $\Lambda \subset M$ be a hyperbolic set of f. Then $E^s(x)$ and $E^u(x)$ vary continuously in $x \in \Lambda$. In particular, $\dim E^s(x)$ and $\dim E^u(x)$ are locally constant. Moreover, the closure $\overline{\Lambda}$ is a hyperbolic set of f.*

Proof. Let $x \in \Lambda$. We prove E^s is continuous at x. It suffices to prove that whenever a sequence $E^s(x_k)$, $x_k \in \Lambda$, converges to a linear subspace $G^s(x)$ of T_xM, then $G^s(x) = E^s(x)$ (because if E^s is not continuous at x, then, taking subsequences, there would be a sequence $x_k \to x$ such that $E^s(x_k)$ converges to a linear subspace $H^s(x)$ of T_xM that is different from $E^s(x)$).

Take any $v \in G^s(x)$. There are $v_k \in E^s(x_k)$ with $v_k \to v$. Since $x_k \in \Lambda$,

$$|Tf^n(v_k)| \le C\lambda^n |v_k|$$

for all $n \ge 0$ and $k \ge 1$. Fixing n and letting $k \to \infty$ gives

$$|Tf^n(v)| \le C\lambda^n |v|,\ \forall\, n \ge 0.$$

By Theorem 4.2, $G^s(x) \subset E^s(x)$. Taking a subsequence if necessary, we may assume that $E^u(x_k)$ converges to a linear subspace $G^u(x)$ of T_xM, and the same proof shows that $G^u(x) \subset E^u(x)$. Since $G^s(x)$ and $G^u(x)$ have complementary dimensions, $G^s(x) = E^s(x)$ (and $G^u(x) = E^u(x)$). This proves $E^s(x)$ varies continuously in x.

Next we prove that $\overline{\Lambda}$ is hyperbolic. It suffices to prove that $\overline{\Lambda} - \Lambda$ is hyperbolic. Let $x \in \overline{\Lambda} - \Lambda$. This time there is not yet a direct sum $E^s(x) \oplus E^u(x)$ given at x. We first construct a direct sum at x. Take a sequence $x_k \in \Lambda$ such that $E^s(x_k)$ and $E^u(x_k)$ converge, respectively, to two linear subspaces $G^s(x)$ and $G^u(x)$ of T_xM. The same argument as above gives

$$|Tf^n(v)| \le C\lambda^n |v|,\ \forall\, v \in G^s(x),\ n \ge 0,$$

$$|Tf^n(v)| \ge C^{-1}(\lambda^{-1})^n |v|,\ \forall\, v \in G^u(x),\ n \ge 0.$$

Then $G^s(x) \cap G^u(x) = \{0\}$. Since $G^s(x)$ and $G^u(x)$ have complementary dimensions, $T_xM = G^s(x) \oplus G^u(x)$. This gives a direct sum at x.

Now we construct an invariant direct sum on $\mathrm{Orb}(x)$. Taking one iterate by Tf, $E^s(fx_k)$ and $E^u(fx_k)$ converge, respectively, to two linear subspaces

$$G^s(fx) = Tf(G^s(x)), \quad G^u(fx) = Tf(G^u(x))$$

of $T_{fx}M$. Since a linear isomorphism preserves direct sum,

$$T_{fx}M = G^s(fx) \oplus G^u(fx).$$

Since the two constants C and λ are independent of points of Λ, vectors of $G^s(fx)$ and $G^u(fx)$ satisfy the same inequalities as vectors of $G^s(x)$ and $G^u(x)$. Taking all positive and negative iterates this way gives an invariant splitting on $\mathrm{Orb}(x)$. Similarly we obtain an invariant splitting for every orbit of $\overline{\Lambda} - \Lambda$. This is obviously a hyperbolic splitting on $\overline{\Lambda} - \Lambda$, proving Theorem 4.3. □

Let $\Lambda \subset M$ be a set. Assume that, for every $x \in \Lambda$, we are given a linear subspace $E(x) \subset T_xM$. As usual, we call

$$E = \bigcup_{x\in\Lambda} E(x)$$

an m-dimensional C^r *subbundle* of $T_\Lambda M$, or an m-dimensional C^r *distribution* on Λ if, for every $x \in \Lambda$, there is a neighborhood U of x in Λ together with m linearly independent C^r vector fields $e_1, \ldots, e_m$ on U such that, for every $y \in U$, the vectors $e_1(y), \ldots, e_m(y)$ span $E(y)$. In this case $E(x)$ is called the *fiber* of E at x. Two C^0 subbundles E_1 and E_2 of $T_\Lambda M$ are said to form a *direct sum*, or *Whitney sum*, denoted $E_1 \oplus E_2$, if $E_1(x)$ and $E_2(x)$ form a direct sum $E_1(x) \oplus E_2(x)$ at every $x \in \Lambda$.

It can be proved that E is a C^0 subbundle of $T_\Lambda M$ if and only if $E(x)$ varies continuously in $x \in \Lambda$ (Exercise 4.5). Thus, by Theorem 4.3, if Λ is hyperbolic, then

$$E^s = \bigcup_{x\in\Lambda} E^s(x), \quad E^u = \bigcup_{x\in\Lambda} E^u(x)$$

are C^0 subbundles of $T_\Lambda M$. It is striking that, in general, the two subbundles E^s and E^u of a hyperbolic set are merely C^0 but not C^1, even if f is C^r with r very large. See Anosov (1967) and Hirsch-Pugh-Shub (1977). This gives a special C^0 flavor for the theory of hyperbolic sets.

Let $\langle\cdot,\cdot\rangle_x$ be an inner product on T_xM, $x \in M$. These inner products put together give a *Riemannian metric* $\langle\cdot,\cdot\rangle$ of TM. We say $\langle\cdot,\cdot\rangle$ is C^r if, acting on every pair of C^∞ local vector fields, $\langle\cdot,\cdot\rangle$ gives a C^r function. Likewise, let $|\cdot|_x$ be a norm on T_xM, $x \in M$. These norms put together give a *norm*, or a *Finsler structure*, of TM. We say $|\cdot|$ is C^r if, acting on every C^∞ local vector field, $|\cdot|^2$ gives a C^r function. A norm that is induced from a Riemannian metric is called a *Riemannian norm*.

As usual, two C^0 norms $|\cdot|$ and $\|\cdot\|$ of TM are called *equivalent* if there is a constant $K \geq 1$ such that, for every $v \in TM$,

$$K^{-1}|v| \leq \|v\| \leq K|v|.$$

We call K a *relative constant* between $|\cdot|$ and $\|\cdot\|$. On a compact manifold all C^0 norms are mutually equivalent.

The next theorem corresponds to Theorem 2.3.

Theorem 4.4. *Let $\Lambda \subset M$ be a hyperbolic set of f with splitting $T_\Lambda M = E^s \oplus E^u$. There are a C^∞ Riemannian metric $\langle\langle \cdot, \cdot \rangle\rangle$ of M and a constant $0 < \tau < 1$ such that, with respect to the induced norm,*

$$\|Tf(v)\| \leq \tau\|v\|, \ \forall \ v \in E^s,$$
$$\|Tf^{-1}(v)\| \leq \tau\|v\|, \ \forall \ v \in E^u.$$

Briefly, there is a Riemannian norm that makes the hyperbolic behavior an immediate contraction and expansion.

Proof. Let $\langle \cdot, \cdot \rangle$ be the given Riemannian metric of M. Take N sufficiently large such that $C\lambda^N < 1$, and define

$$\langle\langle v, u \rangle\rangle = \sum_{n=0}^{N-1} \langle Tf^n(v), Tf^n(u) \rangle, \quad v, u \in T_xM, \ x \in M.$$

Then $\langle\langle \cdot, \cdot \rangle\rangle$ is a Riemannian metric of M. Let $a = \sum_{n=0}^{N-1} C^2\lambda^{2n}$. Then

$$\|v\|^2 \leq a|v|^2, \ \forall \ v \in E^s,$$
$$\|v\|^2 \leq a|Tf^{N-1}(v)|^2, \ \forall \ v \in E^u.$$

We check that $\|\cdot\|$ satisfies the two inequalities of the theorem. In fact, for every $v \in E^s$,

$$\begin{aligned}\|Tf(v)\|^2 &= \|v\|^2 - |v|^2 + |Tf^N(v)|^2 \\ &\leq \|v\|^2 - (1 - C^2\lambda^{2N})|v|^2 \leq \|v\|^2 - a^{-1}(1 - C^2\lambda^{2N})\|v\|^2.\end{aligned}$$

Likewise, for every $v \in E^u$,

$$\begin{aligned}\|Tf^{-1}(v)\|^2 &= \|v\|^2 + |Tf^{-1}(v)|^2 - |Tf^{N-1}(v)|^2 \\ &\leq \|v\|^2 - (1 - C^2\lambda^{2N})|Tf^{N-1}(v)|^2 \leq \|v\|^2 - a^{-1}(1 - C^2\lambda^{2N})\|v\|^2.\end{aligned}$$

Let

$$\tau' = \sqrt{1 - a^{-1}(1 - C^2\lambda^{2N})}.$$

It suffices to verify $0 < \tau' < 1$. But this is obvious because $a \geq 1$.

This Riemannian metric $\langle\langle \cdot, \cdot \rangle\rangle$ is generally C^0 only, because Tf is generally C^0. Take a C^∞ approximation; one obtains a C^∞ Riemannian metric of

M. The induced norm restricted to E^s and E^u satisfies the two inequalities for some $\tau \in (\tau', 1)$. This proves Theorem 4.4. □

A (C^0, not necessarily C^∞) Riemannian metric $\langle\langle \cdot, \cdot \rangle\rangle$ of M that satisfies the two immediate inequalities of Theorem 4.4 is called *adapted* to Λ. We call

$$\tau(\Lambda) = \sup_{x \in \Lambda}\{\|Tf|_{E^s(x)}\|, \ \|Tf^{-1}|_{E^u(x)}\|\} < 1$$

the *skewness* of Λ with respect to the induced norm $\|\cdot\|$.

Let $U \subset M$ be a set, not necessarily invariant, and let $T_U M = E_1 \oplus E_2$ be a direct sum. A norm $|\cdot|$ of $T_U M$ is of *box type* with respect to $E_1 \oplus E_2$ if

$$|v| = \max\{|v_1|, |v_2|\}, \ \forall \ v \in T_U M,$$

where v_1 and v_2 are the two components of v with respect to $E_1 \oplus E_2$. For any norm $|\cdot|$ of $T_U M$, letting

$$\|v\| = \max\{|v_1|, |v_2|\}, \ \forall \ v \in T_U M$$

defines a norm $\|\cdot\|$ on $T_U M$ that is of box type with respect to $E_1 \oplus E_2$, called the *box-adjusted norm* of $|\cdot|$ with respect to $E_1 \oplus E_2$. The box-adjusted norm of an adapted norm to a hyperbolic set Λ with respect to the hyperbolic splitting $E^s \oplus E^u$ of Λ is both adapted to and of box type to Λ with the same skewness.

4.2. Persistence of hyperbolicity for an invariant set

First we fix some standard definitions which are important to this text. Let $\Lambda \subset M$ be f-invariant. Let E_1, E_2 be two C^0 subbundles of $T_\Lambda M$. A map $F : E_1 \to E_2$ is called *fiber-preserving* with respect to f, or is simply said to *cover* f, if

$$\pi F = f \pi,$$

where $\pi : TM \to M$ is the bundle projection. Thus for any $x \in \Lambda$, F maps $E_1(x)$ into $E_2(fx)$. The sum $F + G$ of two fiber-preserving maps F and G that cover the same f is defined in the obvious pointwise way and hence is fiber-preserving over f. Likewise, a fiber-preserving map F over f composed with a fiber-preserving map G over f^{-1} is a fiber-preserving map GF over id.

Let $F : E_1 \to E_2$ be continuous and fiber-preserving over f. In case $F|_{E_1(x)}$ is linear for every $x \in \Lambda$, we call F a C^0 *bundle homomorphism*. Denote

$$|F| = \sup\{|F(v)| \ | \ v \in E_1, \ |v| = 1\}.$$

We say F is *bounded* if $|F| < \infty$. If Λ is compact, a C^0 bundle homomorphism is always bounded. Denote by $L(E_1, E_2; f)$ the set of bounded C^0 bundle homomorphisms from E_1 to E_2 that cover f. With respect to this norm, $L(E_1, E_2; f)$ is a Banach space. If $F|_{E_1(x)}$ is a linear isomorphism for every $x \in \Lambda$, we will say F is a *bundle isomorphism* that covers f. For instance, since f is a diffeomorphism, the tangent map $Tf : T_\Lambda M \to T_\Lambda M$ is a C^0 bundle isomorphism that covers f.

Below in Section 4.4 we will see an important fiber-preserving map which is not linear along fibers.

Remark. The notion of fiber-preserving map is crucial to our text. It gives a suitable setting for proofs about hyperbolic sets so that they look like copies of the corresponding proofs about hyperbolic fixed points. For instance, using the notion of bundle isomorphism, the proof of Lemma 4.5 below will look like a duplicate of the proof of Lemma 2.9. The same role will be played later by a fiber-preserving map that is nonlinear on fibers when we prove the stable manifold theorem and the structural stability theorem for hyperbolic sets.

Now we study the persistence of the hyperbolicity of a hyperbolic set.

Let $T_\Lambda M = E_1 \oplus E_2$ be a C^0 direct sum. Denote by $L(E_1, E_2; id)$ the Banach space of bounded C^0 bundle homomorphisms from E_1 to E_2 that cover id, and denote by $L(E_1, E_2; id)(1)$ the closed unit ball about the origin. If $P \in L(E_1, E_2; id)$, then

$$\mathrm{gr}(P) = \bigcup_{x \in \Lambda} \mathrm{gr}(P_x)$$

is a C^0 subbundle of $T_\Lambda M$.

The next lemma corresponds to Lemma 2.9.

Lemma 4.5. *Let $g : M \to M$ be a diffeomorphism, let Δ be an invariant set of g, and let $B : T_\Delta M \to T_\Delta M$ be a bounded C^0 bundle isomorphism over g, represented under a C^0 direct sum $T_\Delta M = E_1 \oplus E_2$ as*

$$\begin{pmatrix} B_{11} & B_{12} \\ B_{21} & B_{22} \end{pmatrix},$$

where $B_{ij} = \pi_i \circ B|_{E_j}$. If there are a C^0 norm $|\cdot|$ of $T_\Delta M$ that is of box type with respect to $E_1 \oplus E_2$ and two constants $\lambda > 0$ and $\epsilon > 0$ such that

$$\max\{|B_{11}^{-1}|, |B_{22}|\} < \lambda,$$

$$\max\{|B_{12}|, |B_{21}|\} < \epsilon,$$

$$\lambda + \epsilon < 1,$$

then there is a unique C^0 bundle homomorphism $P = P_B : E_1 \to E_2$ over id, $|P| \le 1$, such that the C^0 subbundle $\mathrm{gr}(P)$ is B-invariant; namely, $B_x(\mathrm{gr}(P_x)) = \mathrm{gr}(P_{gx})\ \forall\ x \in \Delta$, and $B_x|_{\mathrm{gr}(P_x)}$ is $(\lambda^{-1} - \epsilon)$-expanding. Moreover, P_B, and hence $\mathrm{gr}(P_B)$, depends continuously on B.

Remark. Since B is fiber-preserving over g, B_{12}, for instance, is automatically from $E_2(x)$ to $E_1(gx)$.

Proof. Since for fiber-preserving maps things are defined pointwise, the proof will be the same as that of Lemma 2.9, except marking base points x. It is strongly recommended that the reader compare the two proofs, sentence by sentence.

Let $P : E_1 \to E_2$ be a C^0 bundle homomorphism over id with $|P| \le 1$. For any $x \in \Delta$ and $v \in E_1(x)$,

$$\begin{pmatrix} B_{11} & B_{12} \\ B_{21} & B_{22} \end{pmatrix}_x \begin{pmatrix} v \\ P_x v \end{pmatrix} = \begin{pmatrix} (B_{11})_x v + (B_{12})_x P_x v \\ (B_{21})_x v + (B_{22})_x P_x v \end{pmatrix}.$$

Hence

$$B_x(\mathrm{gr}(P_x)) \subset \mathrm{gr}(P_{gx}), \ \forall\, x \in \Delta$$

if and only if

$$P_{gx}((B_{11})_x v + (B_{12})_x P_x v) = (B_{21})_x v + (B_{22})_x P_x v, \ \forall x \in \Delta, \ \forall v \in E_1(x);$$

that is,

$$P_{gx}((B_{11})_x + (B_{12})_x P_x) = (B_{21})_x + (B_{22})_x P_x, \ \forall x \in \Delta.$$

Since

$$m((B_{11})_x) \ge \lambda^{-1}, \ \ |(B_{12})_x P_x \le \epsilon,$$

by Theorem 2.7,

$$(B_{11})_x + (B_{12})_x P_x : E_1(x) \to E_1(gx)$$

is invertible. Hence

$$P_{gx} = ((B_{21})_x + (B_{22})_x P_x)((B_{11})_x + (B_{12})_x P_x)^{-1}, \ \forall x \in \Delta.$$

This suggests a map

$$T = T_B : L(E_1, E_2; id)(1) \to L(E_1, E_2; id),$$

$$(T(P))_{gx} = ((B_{21})_x + (B_{22})_x P_x)((B_{11})_x + (E_{12})_x P_x)^{-1}, \ \forall\, x \in \Delta,$$

naturally called the *graph transform* induced by B. Finding a bundle homomorphism P with $B(\mathrm{gr}(P)) \subset \mathrm{gr}(P)$ then reduces to finding a fixed point of T.

We verify that T maps $L(E_1, E_2; id)(1)$ into itself and is a contraction. In fact, for any $P \in L(E_1, E_2; id)(1)$ and any $x \in \Delta$,

$$\begin{aligned} |(T(P))_{gx}| &\le |(B_{21})_x + (B_{22})_x P_x| \cdot |((B_{11})_x + (E_{12})_x P_x)^{-1}| \\ &\le \frac{\lambda + \epsilon}{\lambda^{-1} - \epsilon} < 1. \end{aligned}$$

Hence T maps $L(E_1, E_2; id)(1)$ into itself. Moreover, for any P, $P' \in L(E_1, E_2; id)(1)$ and any $x \in \Delta$,

$$(T(P))_{gx}((B_{11})_x + (B_{12})_x P_x) = (B_{21})_x + (B_{22})_x P_x,$$
$$(T(P'))_{gx}((B_{11})_x + (B_{12})_x P'_x) = (B_{21})_x + (B_{22})_x P'_x.$$

Hence

$$\begin{aligned}((T(P))_{gx} &- (T(P'))_{gx})(B_{11})_x + (T(P))_{gx}(B_{12})_x P_x \\ &- (T(P'))_{gx}(B_{12})_x P_x + (T(P'))_{gx}(B_{12})_x P_x - (T(P'))_{gx}(B_{12})_x P'_x \\ &= (B_{22})_x (P_x - P'_x),\end{aligned}$$

$$\begin{aligned}((T(P))_{gx} &- (T(P'))_{gx})((B_{11})_x + (B_{12})_x P_x) \\ &= ((B_{22})_x - (T(P'))_{gx}(B_{12})_x)(P_x - P'_x),\end{aligned}$$

$$\begin{aligned}(T(P))_{gx} &- (T(P'))_{gx} \\ &= ((B_{22})_x - (T(P'))_{gx}(B_{12})_x)(P_x - P'_x)((B_{11})_x + (B_{12})_x P_x)^{-1}.\end{aligned}$$

Hence

$$|(T(P))_{gx} - (T(P'))_{gx}| \leq \frac{\lambda + \epsilon}{\lambda^{-1} - \epsilon} |P_x - P'_x|, \ \forall x \in \Delta.$$

That is,

$$|T(P) - T(P')| \leq \frac{\lambda + \epsilon}{\lambda^{-1} - \epsilon} |P - P'|.$$

Thus $T = T_B$ is a contraction. By the contraction mapping principle, T has a unique fixed point $P = P_B \in L(E_1, E_2; id)(1)$. In other words, there is a unique $P = P_B \in L(E_1, E_2; id)(1)$ such that

$$B_x(\mathrm{gr}(P_x)) \subset \mathrm{gr}(P_{gx}), \ \forall \ x \in \Delta.$$

That is, $B(\mathrm{gr}(P)) \subset \mathrm{gr}(P)$. Since $B : T_\Delta M \to T_\Delta M$ restricted to every fiber is a linear isomorphism, the inclusion is actually an equality

$$B(\mathrm{gr}(P)) = \mathrm{gr}(P).$$

Since the norm is of box type and since $P \in L(E_1, E_2; id)(1)$, the norm of a vector in $\mathrm{gr}(P)$ is given by the first component. Then

$$|B_x(v, P_x v)| = |(B_{11})_x v + (B_{12})_x P_x v| \geq (\lambda^{-1} - \epsilon)|v|;$$

that is, $B_x|_{\mathrm{gr}(P_x)}$ is $(\lambda^{-1} - \epsilon)$-expanding.

Let $\mathcal{B}$ denote the set of C^0 bundle isomorphisms that satisfy the assumptions of Lemma 4.5. What is discussed above is a family of contractions

$$T : \mathcal{B} \times L(E_1, E_2; id)(1) \to L(E_1, E_2; id)(1)$$
$$T(B, P) = T_B(P) = (B_{21} + B_{22}P)(B_{11} + B_{12}P)^{-1}$$

with parameter B. Clearly T is continuous. As the above computation shows, the contraction rate of T_B is independent of $B \in \mathcal{B}$. By Theorem

2.8, the fixed point P_B, hence the graph $\mathrm{gr}(P_B)$, varies continuously in B. This proves Lemma 4.5. $\square$

Remark. The proof of Lemma 4.5 is the same as Lemma 2.9, just with base points marked. Note that to mark base points is not really necessary because, as a fiber-preserving map, B determines base points automatically. We could have omitted all the base points. But then the proof of Lemma 4.5 would look identical to the proof of Lemma 2.9 and hence could be entirely omitted. This is of course correct mathematically. Nevertheless as one of the main results of this chapter we have still given the proof as above. We will see that most proofs in this chapter can be treated in this fashion, including the two major ones, the stable manifold theorem and the structural stability theorem for hyperbolic sets.

Let $\mathrm{Diff}^r(M)$ be the set of C^r diffeomorphisms of M, endowed with the C^r topology. As usual, denote

$$d(x, \Lambda) = \inf\{d(x, y) \mid y \in \Lambda\}$$

and

$$B(\Lambda, a) = \{x \in M \mid d(x, \Lambda) \leq a\}.$$

The next theorem corresponds to Theorem 2.10.

Theorem 4.6 (Persistence of hyperbolicity for an invariant set). *Let $\Lambda \subset M$ be a compact hyperbolic set of f. There are a C^1 neighborhood $\mathcal{U}_0$ of f in $\mathrm{Diff}^1(M)$ and a number $a_0 > 0$ such that for any $g \in \mathcal{U}_0$, every compact g-invariant set Δ which is contained in $B(\Lambda, a_0)$ is hyperbolic. Moreover, as g C^1 approaches f and $x \in \Delta$ approaches $y \in \Lambda$, the stable subspace $E^s(x, g)$ approaches the stable subspace $E^s(y, f)$. Likewise for the unstable subspaces.*

Proof. Let $T_\Lambda M = E^s \oplus E^u$ be the hyperbolic splitting of Λ. We may assume that the given Riemannian norm $|\cdot|$ of M is adapted to Λ. Since Λ is compact, the C^0 splitting $T_\Lambda M = E^s \oplus E^u$ extends to a (not necessarily invariant) C^0 splitting $T_U M = G^s \oplus G^u$ on a neighborhood U of Λ, meaning that if $x \in \Lambda$, then $E^s(x) = G^s(x)$ and $E^u(x) = G^u(x)$. Let $\|\cdot\|$ be the box-adjusted norm of $|\cdot|$ with respect to $G^s \oplus G^u$. Then $\|\cdot\|$ is defined on $T_U M$ and is both adapted and of box type with respect to $T_\Lambda M = E^s \oplus E^u$.

Let $0 < \tau < 1$ be the skewness of Λ with respect to $\|\cdot\|$ (which equals its skewness with respect to $|\cdot|$). Then Tf on $T_\Lambda M$ is represented as

$$\begin{pmatrix} (Tf)_{uu} & 0 \\ 0 & (Tf)_{ss} \end{pmatrix}$$

with

$$\|(Tf)_{uu}^{-1}\| \leq \tau, \quad \|(Tf)_{ss}\| \leq \tau.$$

Take $\tau < \lambda < 1$ and $\epsilon > 0$ such that $\lambda + \epsilon < 1$. If $\mathcal{U}_0$ and $a_0 > 0$ are sufficiently small, then for any invariant set $\Delta \subset B(\Lambda, a_0) \subset U$ of any $g \in \mathcal{U}_0$, the four block bundle homomorphisms of Tg and Tg^{-1} represented under $T_\Delta M = G^u \oplus G^s|_\Delta$, with respect to $\|\cdot\|$, all satisfy the conditions of Lemma 4.5. The rest of the proof is similar to Theorem 2.10 and hence omitted. □

A norm of box type with respect to a C^0 direct sum $T_\Lambda M = E_1 \oplus E_2$ is generally C^0 only and generally cannot be defined on the whole manifold and must not be induced from any Riemannian metric (violating the parallelogram law). This makes the role played by a norm of box type somewhat different from Chapter 2. In fact, in this chapter we will use a norm of box type only in the middle part of a proof as a supplementary tool. Thus we need to control the relative constant between a norm of box type and the original Riemannian norm. This concerns the angle of a direct sum.

For $x \in \Lambda$, define the *angle* of $E_1(x)$ and $E_2(x)$ to be

$$\angle(E_1(x), E_2(x)) = \inf\{\angle(u, v) \mid u \in E_1(x) - \{0\}, v \in E_2(x) - \{0\}\}.$$

Define

$$\angle(E_1, E_2) = \inf_{x\in\Lambda}\angle(E_1(x), E_2(x)).$$

Let $\Lambda \subset M$ be a hyperbolic set of f with splitting $T_\Lambda M = E^s \oplus E^u$. By Theorem 4.3, $E^s(x)$ and $E^u(x)$ vary continuously in $x \in \Lambda$. Thus, if Λ is compact, then $\angle(E^s, E^u) > 0$.

Lemma 4.7. *For any $\delta > 0$, there is $K \geq 1$ such that for any Euclidean space E, any direct sum $E = E_1 \oplus E_2$, and any inner product $\langle\cdot,\cdot\rangle$ of E, if $\angle(E_1, E_2) > \delta$, where the angle is with respect to $\langle\cdot,\cdot\rangle$, then the box-adjusted norm of the induced norm $|\cdot|$ of $\langle\cdot,\cdot\rangle$ with respect to $E_1 \oplus E_2$ is equivalent to $|\cdot|$ with relative constant K.*

The proof is elementary and is left as an exercise.

Let $\Lambda \subset M$ be a compact hyperbolic set of f. Denote by $|\cdot|_\Lambda$ the box-adjusted norm of $|\cdot|$ with respect to the hyperbolic splitting $T_\Lambda M = E^s \oplus E^u$ of Λ. Thus $|\cdot|_\Lambda$ is defined, but only on $T_\Lambda M$, not on all of TM. For simplicity the notation $|\cdot|_\Lambda$ does not specify the direct sum $E^s \oplus E^u$, but the precise meaning will be clear through the context.

The next lemma adds some details to Theorem 4.6.

Lemma 4.8. *Let $\Lambda \subset M$ be a compact hyperbolic set of f, and let $|\cdot|$ be a Riemannian norm of M. Then there are a C^1 neighborhood $\mathcal{U}_0$ of f in* $\mathrm{Diff}^1(M)$ *and two numbers $a_0 > 0$ and $K \geq 1$ such that every compact invariant set $\Delta \subset B(\Lambda, a_0)$ of every $g \in \mathcal{U}_0$ is hyperbolic and such that the box-adjusted norm $|\cdot|_\Delta$ of $|\cdot|$ with respect to the hyperbolic splitting of Δ*

is equivalent to $|\cdot|$ *with relative constant* K. *Also, if* $|\cdot|$ *is adapted to* Λ *and* $\tau(\Lambda)$ *is the skewness of* Λ *with respect to* $|\cdot|$, *then for any* $\epsilon > 0$, *there are a* C^1 *neighborhood* $\mathcal{U} \subset \mathcal{U}_0$ *of* f *in* $\mathrm{Diff}^1(M)$ *and a number* $0 < a < a_0$ *such that the skewness with respect to* $|\cdot|$ *of every compact invariant set* $\Delta \subset B(\Lambda, a)$ *of every* $g \in \mathcal{U}$ *satisfies* $\tau(\Delta) \leq \tau(\Lambda) + \epsilon$.

Proof. The proof is standard and we give a sketch only. If g is C^1 close to f and Δ is in a small neighborhood of Λ, then, by Theorem 4.6, for every $x \in \Delta$ there is $y \in \Lambda$ such that $E^s(x,g) \oplus E^u(x,g)$ is close to $E^s(y,f) \oplus E^u(y,f)$. Hence the angle of $E^s(x,g) \oplus E^u(x,g)$ is close to the angle of $E^s(y,f) \oplus E^u(y,f)$. By Lemma 4.7, there is a constant $K \geq 1$ independent of Δ such that $|\cdot|_\Delta$ is equivalent to $|\cdot|$ by the constant K. Also, if g is C^1 close to f and Δ is in a small neighborhood of Λ, then if $|\cdot|$ is adapted to Λ, then it is adapted to Δ. The rest of the proof is obvious. □

4.3. Smoothness in Lemma 2.17 and Theorem 2.18

Our manifold M has always been assumed compact. This is for the general need of the text. To complete the proofs of Lemma 2.17 and Theorem 2.18 we make in this section some comments about hyperbolic sets of a noncompact manifold. We omit the definition which will be the same as the one stated in Section 4.1, with an extra assumption that norms of the tangent maps Tf and Tf^{-1} at all points are bounded above, a condition that is automatically satisfied for compact manifolds.

We consider a very special case only, when the noncompact manifold is a finite-dimensional normed vector space E and the diffeomorphism $f : E \to E$ is of the form $A + \phi$, where $A : E \to E$ is a hyperbolic linear isomorphism and $\phi : E \to E$ is C^1 and has small Lipschitz constant. We can check that Theorems 4.2 and 4.3 and Lemma 4.5 still hold for this setting. In the proof of Theorem 4.3 the Grassmann space $G^m(M)$ is compact. Now $G^m(E)$ is locally compact, which suffices for the proof of Theorem 4.3. An important observation is that $A+\phi$ *will have the whole space* E *as a hyperbolic set. In other words,* $A + \phi$ *will be Anosov.*

We remark that the noncompact setting considered here is specific and limited. In general a noncompact setting could yield very different phenomenon. Warren White (1973) has a complete metric on the plane that makes a translation hyperbolic.

Now we complete the proof of Lemma 2.17. We were left to prove item (2). Recall that E is a finite-dimensional normed vector space, $A : E \to E$ is a hyperbolic linear isomorphism with splitting $E = E^u \oplus E^s$, to which the norm $|\cdot|$ of E is adapted and of box type, and $\phi : E \to E$ is Lipschitz with $\phi(0) = 0$. We proved in item (1) that if $\mathrm{Lip}\,\phi$ is small enough, $W^u(0, A+\phi)$

will be exactly the graph $\mathrm{gr}(\sigma)$ of a Lipschitz map $\sigma : E^u \to E^s$ with $\mathrm{Lip}\,\sigma \leq 1$. Now assume in addition that $\phi : E \to E$ is C^1. We prove that if $\mathrm{Lip}\,\phi$ is small enough, then σ is C^1 and the C^1 submanifold $W^u(0, A+\phi)$ is tangent at the origin to the unstable subspace G^u of the hyperbolic linear isomorphism $A + D\phi(0)$.

Abbreviate

$$g = A + \phi : E \to E.$$

If $\mathrm{Lip}\,\phi$ is sufficiently small, by the remark after Theorem 2.7, g will be a diffeomorphism of E. We prove that if $\mathrm{Lip}\,\phi$ is small, then the whole space E is a hyperbolic set of g; that is, g is Anosov.

First we point out that A itself is Anosov. Thus we regard E as a manifold and A as a diffeomorphism. For every $x \in E$, define

$$E^u(x) = \{x\} \times E^u, \quad E^s(x) = \{x\} \times E^s.$$

Then

$$T_xE = E^u(x) \oplus E^s(x).$$

Since

$$TA(v) = DA(x) \cdot v = Av,$$

where $x = \pi v$ and $\pi : TE \to E$ is the bundle projection, it is easy to see that this is a hyperbolic splitting for A. Thus the whole space E is a hyperbolic set of A; namely A is Anosov.

We check that if $\mathrm{Lip}\,\phi$ is small enough, then g is also Anosov. In fact, with respect to the splitting $E^u(x) \oplus E^s(x)$, $x \in E$, we have

$$Tg = \begin{pmatrix} A_{uu} + (T\phi)_{uu} & (T\phi)_{us} \\ (T\phi)_{su} & A_{ss} + (T\phi)_{ss} \end{pmatrix},$$

where

$$|A_{uu}^{-1}| \leq \tau, \quad |A_{ss}| \leq \tau,$$

and $0 < \tau < 1$ is the skewness of A. If $\mathrm{Lip}\,\phi$ is sufficiently small, $|T\phi|$ will be arbitrarily small; hence Tg and Tg^{-1} will both satisfy Lemma 4.5; hence E will be a hyperbolic set of g. Let

$$T_xE = G^u(x) \oplus G^s(x), \quad x \in E,$$

be the hyperbolic splitting of Tg. Note that

$$\dim G^u(x) = \dim E^u.$$

Now we prove σ is C^1. Take any $v \in E^u$. First we prove σ is differentiable at v. Denote $z = (v, \sigma(v)) \in \mathrm{gr}(\sigma)$ for short. By the criterion of Katok-Hasselblatt (1995) stated at the end of Chapter 2, we need to prove that the tangent set $T_z\mathrm{gr}(\sigma)$ is contained in a linear subspace of T_zE of dimension $\dim E^u$. In fact, we prove

$$T_z\mathrm{gr}(\sigma) \subset G^u(z).$$

Since $\mathrm{Lip}\,\sigma \leq 1$, for every $x \in \mathrm{gr}(\sigma)$, every generalized tangent line of $T_x\mathrm{gr}(\sigma)$ lies in the 1-cone $C_1(E^u(x))$ with respect to the direct sum $E^u(x) \oplus E^s(x)$. As long as $\mathrm{Lip}\,\phi$ is small enough, $G^u(x)$ and $G^s(x)$ will be close to $E^u(x)$ and $E^s(x)$, respectively; hence the generalized tangent lines of $T_x\mathrm{gr}(\sigma)$ will be contained in the 2-cone $C_2(G^u(x))$ with respect to the direct sum $G^u(x) \oplus G^s(x)$. Since $\mathrm{gr}(\sigma)$ is invariant under g, Tg^{-1} maps generalized tangent lines of $\mathrm{gr}(\sigma)$ into generalized tangent lines of $\mathrm{gr}(\sigma)$. In particular, for every generalized tangent line l of $T_z\mathrm{gr}(\sigma)$,

$$Tg^{-n}(l) \subset C_2(G^u(g^{-n}z)),\ \forall n \geq 0.$$

See Figure 4.2. By Theorem 4.2,

$$l \subset G^u(z).$$

This proves $T_z\mathrm{gr}(\sigma) \subset G^u(z)$.

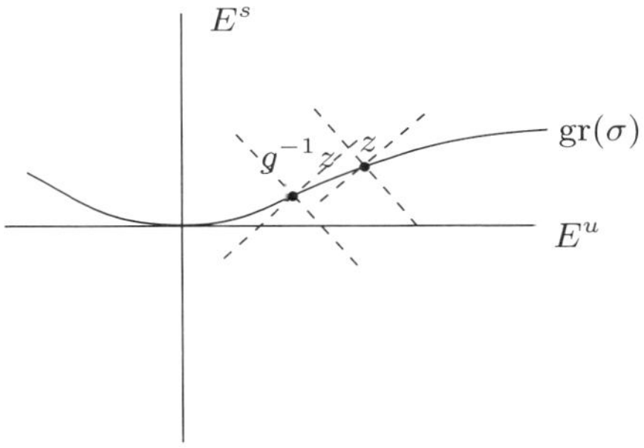

Figure 4.2. For any $x \in \mathrm{gr}(\sigma)$, every generalized tangent line of $\mathrm{gr}(\sigma)$ through x is contained in the cone $C_2(G^u)(x)$. Since the graph $\mathrm{gr}(\sigma)$ is g-invariant, every generalized tangent line of $\mathrm{gr}(\sigma)$ remains in these cones under repeated iterates of Tg^{-1}.

By the criterion, σ is differentiable at v, and $\mathrm{gr}(\sigma)$ is tangent at z to the unstable subspace $G^u(z)$ of Tg. In particular, at the origin, $\mathrm{gr}(\sigma)$ is tangent to the unstable subspace G^u of $A + D\phi(0)$. By Theorem 4.3, $G^u(z)$ varies continuously in $z \in \mathrm{gr}(\sigma)$. Thus σ is C^1. This finishes the proof of item (2) of Lemma 2.17. Thus the proof of Lemma 2.17 is complete. □

Now we prove Theorem 2.18. Let $f : U \to E$ be C^k, $k \geq 1$, and let $0 \in U$ be a hyperbolic fixed point of f with splitting $E = E^s \oplus E^u$. We prove there is $r > 0$ such that $W_r^s(0, f)$ is a C^k embedded submanifold of E of dimension $\dim E^s$, tangent at 0 to E^s.

First we prove the case $k = 1$. Let f be C^1. We first work with a norm $|\cdot|$ of E that is adapted to and of box type to $E^s \oplus E^u$. We extend the locally defined f to the whole E and apply item (2) of Lemma 2.17 to obtain the (global) stable manifold; then we cut it off to get the local stable manifold.

Fix a C^∞ bump function $\alpha : E \to \mathbb{R}$ with $0 \le \alpha \le 1$ such that $\alpha(v) = 1$ for $|v| \le 1/3$ and $\alpha(v) = 0$ for $|v| \ge 2/3$. Abbreviate $A = Df(0)$. Write

$$\phi_f = f - A : U \to E.$$

Then ϕ_f is C^1, and

$$\phi_f(0) = 0, \quad D\phi_f(0) = 0.$$

(Here ϕ_f has an additional property of "tangency" compared to the map ϕ of Lemma 2.17.) Define

$$\overline{\phi}_f : E \to E$$
$$\overline{\phi}_f(v) = \alpha\left(\frac{v}{3r}\right)\phi_f(v),$$

where $r > 0$ satisfies

$$E(3r) \subset U$$

and will be shrunk shortly. Then $\overline{\phi}_f$ is C^1, and $\overline{\phi}_f = \phi_f$ on $E(r)$. By the claim in the proof of Theorem 2.13, if $r > 0$ is sufficiently small, then $\operatorname{Lip}\overline{\phi}_f$ on the whole E, with respect to $|\cdot|$, will be small enough to satisfy Lemmas 2.14 and 2.17. (In particular, $\overline{f} = A + \overline{\phi}_f$ will be Anosov. Thus, on a neighborhood of a hyperbolic fixed point, f always agrees with some Anosov diffeomorphism $\overline{f}$.) By Lemma 2.17, there is a C^1 map

$$\sigma : E^s \to E^u$$

with $\sigma(0) = 0$ and $\operatorname{Lip}\sigma \le 1$ such that

$$W^s(0, A + \overline{\phi}_f) = \operatorname{gr}(\sigma).$$

Moreover, the C^1 submanifold $W^s(0, A + \overline{\phi}_f)$ is tangent at $0 \in E$ to the stable subspace of the hyperbolic linear isomorphism $A + D\phi_f(0)$. Since

$$D\phi_f(0) = 0,$$

this stable subspace is just E^s. This means

$$D\sigma(0) = 0.$$

Since the norm $|\cdot|$ is of box type,

$$E(r) = E^s(r) \times E^u(r).$$

Denote

$$i : E^s \to E$$

to be

$$i(v) = (v, \sigma(v)).$$

Then i is a C^1 embedding that takes E^s onto the C^1 submanifold $\mathrm{gr}(\sigma)$, tangent at $0 \in E$ to E^s. Since $\operatorname{Lip} \sigma \le 1$, we have

$$i(E^s(r)) = W^s(0, A + \overline{\phi}_f) \cap E(r).$$

We prove

$$W_r^s(0, f) = W^s(0, A + \overline{\phi}_f) \cap E(r).$$

Since $\overline{\phi}_f = \phi_f$ on $E(r)$, the "$\subset$" part is obvious. We prove "$\supset$". Let $v \in W^s(0, A + \overline{\phi}_f) \cap E(r)$. It suffices to prove $(A + \overline{\phi}_f)^n v \in E(r)$ for all $n \ge 1$. But this is obvious because, by Lemma 2.14 (for the case of the whole space E),

$$W^s(0, A + \overline{\phi}_f) = \{v \in E \mid |(A + \overline{\phi}_f)^n v| \le (\tau + \operatorname{Lip} \overline{\phi}_f)^n |v|, \ \forall n \ge 0\}.$$

This proves "$\supset$". Thus

$$W_r^s(0, f) = i(E^s(r))$$

is a C^1 submanifold of E.

So far we have worked with a norm $|\cdot|$ of E that is adapted to and of box type to $E^s \oplus E^u$. Now we get back to the original norm $\|\cdot\|$ of E. Take $0 < a < r$ such that

$$W_a^s(0, f; \|\cdot\|) \subset W_r^s(0, f; |\cdot|).$$

Then $W_a^s(0, f; |\cdot\|)$ is a C^1 submanifold of E. Let

$$V = i^{-1}(W_a^s(0, f; \|\cdot\|)).$$

Then V is a neighborhood of 0 in E^s that is mapped onto $W_a^s(0, f; \|\cdot\|)$ by the graph of σ. This proves the case $k = 1$.

Now we prove the case $k \ge 2$. Since the tangent plane has been determined in the case $k = 1$, we prove the smoothness of $W_r^s(0, f)$ only. The proof is taken from Robinson (1995). We use induction. Assume the case of $k - 1$ has been proved; we prove the case of k.

Let f be C^k. Consider the map

$$F : U \times E \to E \times E$$

$$F(x, v) = (fx, \ Df(x)v).$$

Here the product space takes the usual max metric. Then F is C^{k-1}. Clearly,

$$F^n(x, v) = (f^n(x), \ Df^n(x)v).$$

Also,

$$F(0, 0) = (0, 0),$$

and

$$DF(0, 0) = \begin{pmatrix} Df(0) & 0 \\ 0 & Df(0) \end{pmatrix}.$$

Hence $(0,0)$ is a hyperbolic fixed point of F. By induction, there is $r > 0$ such that $W^s_r((0,0),F)$ is a C^{k-1} submanifold of $U \times E$.

On the other hand, by Theorem 2.15, we may assume $r > 0$ has been chosen such that

$$W^s_r((0,0),F) = \{(x,v) \in U \times E \mid |f^n x| \le r,\ |Df^n(x)v| \le r,\ \forall n \ge 0\},$$

where $Df^n(x)v$ is just $Tf^n(v)$. As proved before (in the case of $k=1$), f always agrees with an Anosov diffeomorphism $\overline{f}$ of E on a neighborhood of a hyperbolic fixed point. Let

$$G^s(x) \oplus G^u(x), \quad x \in E,$$

be the hyperbolic splitting of $\overline{f}$. If $r > 0$ is sufficiently small, then $v \in T_x E$ satisfies

$$|Tf^n(v)| \le r, \quad \forall n \ge 0$$

if and only if

$$v \in G^s(x)(r).$$

Hence the above equation can be rewritten as

$$W^s_r((0,0),F) = \{(x,v) \in U \times E \mid x \in W^s_r(0,f),\ v \in G^s(x)(r)\}.$$

As proved before,

$$G^s(x) = T_x(W^s(0,f)).$$

Hence $W^s_r((0,0),F)$ is just (the r-neighborhood of the 0-set of) the tangent bundle of $W^s_r(0,f)$. Since the degrees of smoothness of a manifold and its tangent bundle differ by 1 and since $W^s_r((0,0),F)$ is C^{k-1}, so $W^s_r(0,f)$ is C^k. This proves the case of k, finishing the proof of Theorem 2.18. □

Summary. The proof of the local stable manifold theorem of a hyperbolic fixed point is long. It goes through Lemma 2.17 and Theorem 2.18 and crosses two chapters. Let us give a summary. The proof consists of four steps. Steps 1 and 2 correspond to Lemma 2.17. The other two steps correspond to Theorem 2.18.

Step 1. $A+\phi$, $\phi : E \to E$ Lipschitz, $\operatorname{Lip}\phi$ small.

We carry out the graph transform to get an invariant graph $\operatorname{gr}(\sigma)$, where $\sigma : E^u \to E^s$ is Lipschitz, $\operatorname{Lip}\sigma \le 1$. We verify $W^u(0, A+\phi) = \operatorname{gr}(\sigma)$.

Step 2. $A+\phi$, $\phi : E \to E$ C^1, $\operatorname{Lip}\phi$ small.

This is the same setting as Step 1, just with ϕ C^1. We verify that $A+\phi$ is Anosov. Using Theorem 4.2 and the criterion on generalized tangent lines, it follows that $W^u(0, A+\phi)$ is C^1.

Step 3. $f : U \to E$ C^1, $0 \in E$ a hyperbolic fixed point.

Write $f = A + \phi_f$, extend ϕ_f to $\overline{\phi}_f$, then apply Step 2 to $A + \overline{\phi}_f$ to get a C^1 map $\sigma : E^s \to E^u$ such that $W^s(0, A + \overline{\phi}_f) = \mathrm{gr}(\sigma)$. Then we cut off the graph to get $W_r^s(0, f)$.

Step 4. $f : U \to E$ C^k, $k \geq 2$, $0 \in E$ a hyperbolic fixed point.

Consider the "tangent map" $F(x, v) = (fx,\ Df(x)v)$. It is C^{k-1} if f is C^k. We verify that $W_r^s((0,0), F)$ is (the r-neighborhood of the 0-set of) the tangent bundle of $W_r^s(0, f)$. Then the proof goes by induction.

4.4. Stable manifolds of hyperbolic sets

Now we come back to our compact manifold M. Taking closure if necessary, we assume that all hyperbolic sets are compact.

In this section we study stable manifolds for hyperbolic sets, a fundamental topic to differentiable dynamical systems. As Smale (1980) commented, the global stable manifolds "lie close to the heart of the subject". The basic references for the stable manifolds theory are Hirsch-Pugh (1970) and Hirsch-Pugh-Shub (1977).

As indicated in the preface, our strategy is to choose a suitable setting so that the proof for hyperbolic sets will match that for hyperbolic fixed points. This is the setting of "fiber-stable manifolds" defined below. First we make some technical preparations.

Recall that a basic approach in Chapter 2 is to take the difference

$$\phi = f - Df(0)$$

and try to get $\mathrm{Lip}\,\phi$ small. However, on a manifold, $f - Tf$ does not make sense in general. What we do is to use the exponential map to "lift" f locally to the tangent bundle TM so that a subtraction will make sense.

Let $x \in M$. Recall that the *exponential map*

$$exp_x : T_xM \to M$$

at $x \in M$ is defined to be

$$exp_x(v) = \sigma_v(1),$$

where $\sigma_v(t)$ is the geodesic determined by the Riemannian metric of M, through x at $t = 0$ with velocity v. (Here M is compact; hence exp_x can be defined on the whole T_xM.) See Figure 4.3.

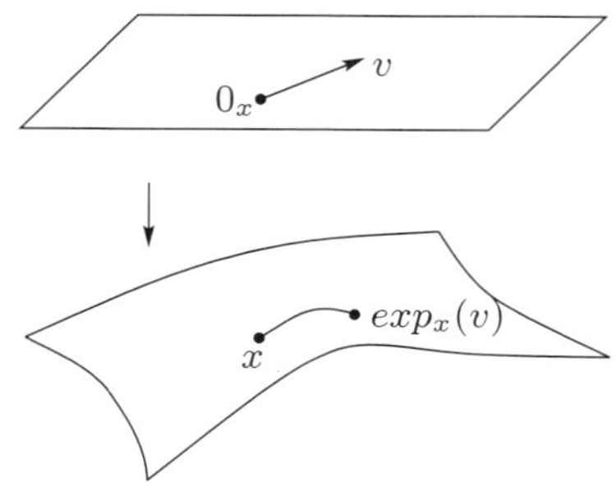

Figure 4.3. The exponential map.

Recall our notation

$$T_xM(\rho) = \{v \in T_xM \mid |v| \leq \rho\}.$$

The following theorem can be found in a general textbook on differential geometry, for instance Boothby (1975). We omit the proof.

Theorem 4.9. (1) *$exp_x(0_x) = x$, where 0_x is the origin of T_xM.*

(2) $D(exp_x)(0_x) : T_xM \to T_xM$ is the identity.

(3) There is $\rho > 0$ such that, for any $x \in M$, $exp_x : T_xM(\rho) \to M$ is a C^∞ embedding. Moreover, $d(x, exp_x(v)) = |v|$, $\forall\ v \in T_xM(\rho)$, where d and $|\cdot|$ are both induced by the given Riemannian metric of M.

(4) The map $exp : TM \to M$, $exp(v) = exp_{\pi v}(v)$, is C^∞, where $\pi : TM \to M$ is the bundle projection.

Thus, taking x as the base, any nearby point $y \in B(x,\rho)$ determines a vector $exp_x^{-1}y \in T_xM$ of length $|exp_x^{-1}y| = d(x,y)$. In a Euclidean space it is just the vector $y - x$ from x to y.

Through the exponential map, f is locally lifted to the tangent bundle to a fiber-preserving map over f, which is not linear on fibers. Precisely, fix $0 < r_\rho < \rho$ such that, for any two points $x, y \in M$, if $d(x,y) < r_\rho$, then $d(fx, fy) < \rho$. Recall by definition

$$TM(r_\rho) = \{v \in TM \mid |v| \leq r_\rho\}.$$

Define the *self-lifting*

$$F_f : TM(r_\rho) \to TM$$

of f to be

$$F_f(v) = exp_{f(x)}^{-1}\ f\ exp_x(v), \quad x = \pi v;$$

see Figure 4.4.

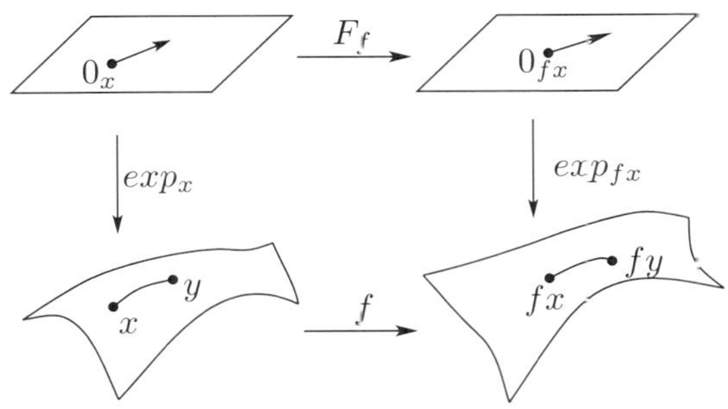

Figure 4.4. The self-lifting F_f of f.

Clearly, F_f is fiber-preserving over f. Since f is C^1, so is F_f. Briefly, thinking of points of an orbit of f as "moving origins", F_f exhibits the behavior of f near these origins. Clearly,

$$F_f(exp_x^{-1}y) = exp_{fx}^{-1}(fy).$$

Thus F_f takes the vector "from x to y" to the vector "from fx to fy". Inductively,

$$F_f^n(exp_x^{-1}y) = exp_{f^n x}^{-1}(f^n y).$$

In particular,

$$|F_f^n(exp_x^{-1}y)| = d(f^n x, f^n y).$$

In this way the distance $d(f^n x, f^n y)$ of two points on M is converted to the length $|exp_{f^n x}^{-1}(f^n y)|$ of a vector on TM. Of course, for these iterates $F_f^n(v)$ to make sense, where $v = exp_x^{-1}y$, we have to assume

$$d(f^i x, f^i y) < r_\rho, \ \forall\, 0 \le i \le n-1.$$

For a vector $v \in TM(r_\rho)$, the iterate $F_f^n(v)$ may not be well defined for all $n \ge 0$. This is as in Chapter 2, where a vector $v \in E(r)$ may not have the iterates $(A+\phi)^n(v)$ contained in $E(r)$ for all $n \ge 0$. The box $E(r)$ was like a window of size r with center $0 \in E$, where we looked at the dynamics of $A+\phi$ for $n \ge 0$. We only considered vectors v whose positive iterates $(A+\phi)^n v$ all remain in the window $E(r)$. Now for every $x \in M$ we have in the tangent plane of x a window $T_xM(r_\rho)$ of center $0_x \in T_xM$ and of size r_ρ, where we look at the dynamics of F_f for $n \ge 0$. The windows are by definition mutually disjoint. We only consider vectors v whose positive iterates $F_f^n(v)$ all remain in the union

$$TM(r_\rho) = \bigcup_{x \in M} T_xM(r_\rho)$$

of the windows. The only difference is that, when we look at the dynamics, we are moving from one window to another.

A fiber-preserving map $F : TM(r) \to TM$ over f, when restricted to every fiber, becomes a map between two Euclidean spaces:

$$F|_{T_xM(r)} : T_xM(r) \to T_{fx}M.$$

Define the *fiber-derivative* of F at $v \in TM(r)$ to be

$$D_2F(v) = D(F|_{T_xM(r)})(v) : T_xM(r) \to T_{fx}M,$$

where $x = \pi v$. In local coordinates, $TM(r)$ is represented as $B_1 \times B_2$, where B_1 and B_2 are balls of $\mathbb{R}^d$, $d = \dim M$, with B_1 representing the base part, and B_2 the fiber part. Thus D_2F is just the partial derivative of F with respect to the second variable. The *higher-order fiber-derivatives* are defined likewise to be the higher-order derivatives of the restricted map.

Lemma 4.10. *Let $f : M \to M$ be a C^1 diffeomorphism.*

(1) $F_f(0_x) = 0_{fx}, \ \forall\, x \in M$.

(2) $D_2(F_f)(0_x) = T_xf, \ \forall\, x \in M$.

(3) $D_2(F_f)$ *is continuous on* $TM(r_\rho)$.

Remark. In local coordinates, item (3) says that the partial derivative of F_f with respect to the second variable is continuous on $B_1 \times B_2$ (with respect to both variables).

Proof.

$$F_f(0_x) = exp_{fx}^{-1}\ f\ exp_x(0_x) = exp_{fx}^{-1}(fx) = 0_{fx}.$$

$$\begin{aligned} D_2(F_f)(0_x) &= D(exp_{fx}^{-1}\ f\ exp_x)(0_x) \\ &= id|_{T_{fx}M} \circ Df(x) \circ id|_{T_xM} = T_xf. \end{aligned}$$

Let $x = \pi v$. Then

$$\begin{aligned} D_2(F_f)(v) &= D(exp_{fx}^{-1}\ f\ exp_x)(v) \\ &= D(exp_{fx}^{-1})(f(exp_x v)) \circ Df(exp_x v) \circ D(exp_x)(v). \end{aligned}$$

Since f is C^1, $D_2(F_f)$ is continuous on $TM(r_\rho)$. □

Let

$$\phi_f = F_f - Tf : TM(r_\rho) \to TM.$$

Then ϕ_f is fiber-preserving for f and, by Lemma 4.10,

$$\phi_f(0_x) = 0_{fx}, \quad D_2\phi_f(0_x) = 0, \quad \forall\, x \in M.$$

Note that ϕ_f is C^0 only (since Tf is C^0 only), though F_f is C^1. However ϕ_f restricted to every fiber $T_xM(r_\rho)$ is C^1. In fact we have more: $D_2\phi_f$ is continuous on $TM(r_\rho)$. In terms of local coordinates, the former means that the partial derivative of ϕ_f with respect to the second variable is continuous with respect to the second variable, and the latter means that the partial derivative of ϕ_f with respect to the second variable is continuous with respect to both variables.

For a continuous fiber-preserving map $F : TM(r) \to TM$ that is Lipschitz on every fiber, define the *fiber-Lipschitz constant* of F to be

$$\mathrm{Lip}_2 F = \sup_{x\in M} \mathrm{Lip}(F|_{T_xM(r)}).$$

This will be the only form of Lipschitz constant below we will consider for a fiber-preserving map.

Lemma 4.11. *Let $f : M \to M$ be a diffeomorphism. Denote $\phi_g = F_g - Tg$. Then for any $\epsilon > 0$, there are a C^1 neighborhood $\mathcal{U}$ of f and a number $r > 0$ such that, for any $g \in \mathcal{U}$, $\mathrm{Lip}_2\phi_g < \epsilon$ on $TM(r)$.*

Proof. Since $D_2\phi_f$ is continuous on $TM(r_\rho)$ and since $D_2\phi_f(0_x) = 0$ and M is compact, for any $\epsilon > 0$, there is $r > 0$ such that for any $v \in TM(r)$ one has $|D_2\phi_f(v)| < \epsilon$. But $TM(r)$ is compact; hence there is a C^1 neighborhood $\mathcal{U}$ of f such that for any $g \in \mathcal{U}$ and any $v \in TM(r)$ one has $|D_2\phi_g(v)| < \epsilon$. Applying the generalized mean value theorem to fibers, we get $\mathrm{Lip}_2\phi_g < \epsilon$ on $TM(r)$. □

Now we study stable manifolds for hyperbolic sets. Since the global stable manifolds are obtained by iterating the local stable manifolds, we start with the local ones.

Recall that, in Chapter 2, the basic setting for the problem of stable manifold is $A + \phi$, where A is a hyperbolic linear isomorphism of a finite-dimensional normed vector space E and ϕ is Lipschitz with Lipschitz constant small. Thus, here we start with maps of the form

$$Tf + \phi$$

on $T_\Lambda M$, where ϕ is continuous, fiber-preserving over f, fiber-Lipschitz with $\mathrm{Lip}_2\phi$ small. It could be referred to as a *fiber-Lipschitz perturbation of a hyperbolic bundle isomorphism.*

Let

$$\phi : TM(r) \to TM, \ 0 < r \le \infty,$$

be fiber-preserving over f such that

$$\phi(0_x) = 0_{fx}, \ \forall\, x \in M.$$

Let $|\cdot|$ be a norm on TM. Define, respectively, the *local fiber-stable manifold* and the *local fiber-unstable manifold* of 0_x of size r with respect to $Tf + \phi$ to be

$$W_r^s(0_x, Tf + \phi) = \{v \in T_xM \mid |(Tf + \phi)^n v| \le r \ \forall n \ge 0, \text{ and } \lim_{n\to+\infty} |(Tf + \phi)^n v| = 0\},$$

$$W_r^u(0_x, Tf + \phi) = \{v \in T_xM \mid |(Tf + \phi)^{-n} v| \le r \ \forall n \ge 0, \text{ and } \lim_{n\to+\infty} |(Tf + \phi)^{-n} v| = 0\}.$$

Here we have used the term "fiber-stable" because v is required to be in the same fiber of 0_x. This will be the only form of stable and unstable manifolds below that we will consider for a fiber-preserving map.

Our setting will be fiber-stable manifolds of a fiber-Lipschitz perturbation of Tf. Within this setting, the proof of the stable manifold theorem of a hyperbolic set will match that of a hyperbolic fixed point, like a copy.

Now we "repeat" what we did in Section 2.5. The next lemma corresponds to Lemma 2.14.

Lemma 4.12 (Characterization of W_r^s on fibers). *Let $\Lambda \subset M$ be a hyperbolic set of f with splitting $T_\Lambda M = E^s \oplus E^u$ of skewness $0 < \tau < 1$ with respect to a C^0 norm $|\cdot|$ of $T_\Lambda M$ that is adapted to and of box type to $E^s \oplus E^u$. Let $r > 0$. Let $\phi : T_\Lambda M(r) \to T_\Lambda M$ be continuous, fiber-preserving over f, and fiber-Lipschitz such that*

$$\mathrm{Lip}_2\phi < 1 - \tau, \quad \phi(0_x) = 0_{fx} \quad \forall x \in \Lambda.$$

Then for any $x \in \Lambda$,

$$\begin{aligned} W_r^s(0_x, Tf + \phi) &= \{v \in T_xM(r) \mid |(Tf + \phi)^n v| \le r, \ \forall n \ge 0\} \\ &= \{v \in T_xM(r) \mid (Tf + \phi)^n v \in T_{f^n x}M(r) \cap C_1(E^s(f^n x)), \ \forall n \ge 0\} \\ &= \{v \in T_xM(r) \mid |(Tf + \phi)^n v| \le (\tau + \mathrm{Lip}_2\phi)^n |v|, \ \forall n \ge 0\}. \end{aligned}$$

Likewise for W_r^u.

Proof. Here $T_\Lambda M(r)$, Tf, ϕ, and $\mathrm{Lip}_2\phi$ correspond to $E(r)$, A, ϕ, and $\mathrm{Lip}\,\phi$ of Lemma 2.14. We write the first part of the proof only.

Claim 1. *If $x \in \Lambda$ and $v, v' \in T_xM(r)$, then*

$$|(Tf + \phi)_s(v) - (Tf + \phi)_s(v')| \le (\tau + \mathrm{Lip}_2\phi)|v - v'|.$$

In fact,

$$\begin{aligned} |(Tf + \phi)_s(v) - (Tf + \phi)_s(v')| &= |(Tf)_{ss}(v_s - v'_s) + \phi_s(v) - \phi_s(v')| \\ &\le (\tau + \mathrm{Lip}_2\phi)|v - v'|. \end{aligned}$$

Claim 2. *If $v, v' \in T_xM(r)$ and $v - v' \notin C_1(E^s(x))$, then*

$$(Tf+\phi)v - (Tf+\phi)v' \notin C_1(E^s(fx))$$

and

$$|(Tf+\phi)_u v - (Tf+\phi)_u v'| \geq (\tau^{-1} - \mathrm{Lip}_2\phi)|v - v'|.$$

In fact,

$$\begin{aligned}|(Tf+\phi)_u v - (Tf+\phi)_u v'| &= |(Tf)_{uu}(v_u - v'_u) + \phi_u(v) - \phi_u(v')| \\ &\geq \tau^{-1}|v_u - v'_u| - \mathrm{Lip}_2\phi\, v - v'|.\end{aligned}$$

But $v - v' \notin C_1(E^s(x))$; hence $|v_u - v'_u| = |v - v'|$. Then

$$|(Tf+\phi)_u v - (Tf+\phi)_u v'| \geq (\tau^{-1} - \mathrm{Lip}_2\phi)|v - v'|.$$

Now $v - v' \notin C_1(E^s(x))$; hence $v - v' \neq 0$. Combined with Claim 1 we get

$$|(Tf+\phi)_u v - (Tf+\phi)_u v'| > |(Tf+\phi)_s v - (Tf+\phi)_s v'|.$$

Thus $(Tf+\phi)v - (Tf+\phi)v' \notin C_1(E^s(fx))$. This proves Claim 2.

Clearly, the proof is a duplicate of that of Lemma 2.14; hence we stop here. □

Remark. As remarked after Lemma 2.14, here $T_\Lambda M(r)$ could be the whole $T_\Lambda M$.

Now we pass to our manifold M. For any $x \in M$ and $r > 0$, define the *local stable manifold* and *local unstable manifold* of x of size r with respect to f to be, respectively,

$$\begin{aligned}&W_r^s(x, f) \\ &\quad = \{y \in M \mid d(f^n y, f^n x) \leq r\ \forall\, n \geq 0, \text{ and } \lim_{n\to+\infty} d(f^n y, f^n x) = 0\},\end{aligned}$$

$$\begin{aligned}&W_r^u(x, f) \\ &\quad = \{y \in M \mid d(f^{-n} y, f^{-n} x) \leq r,\ \forall\, n \geq 0, \text{ and } \lim_{n\to+\infty} d(f^{-n} y, f^{-n} x) = 0\}.\end{aligned}$$

Clearly, for any $x \in M$,

$$f(W_r^s(x)) \subset W_r^s(fx), \quad f(W_r^u(x)) \supset W_r^u(fx).$$

The next theorem corresponds to Theorem 2.15.

Theorem 4.13 (Characterization of W_r^s on manifold). *Let $\Lambda \subset M$ be a hyperbolic set of f. There are $r > 0$, $C \geq 1$, and $0 < \lambda < 1$ such that for any $x \in \Lambda$,*

$$\begin{aligned}W_r^s(x, f) &= \{y \in M \mid d(f^n y, f^n x) \leq r,\ \forall\, n \geq 0\} \\ &= \{y \in M \mid d(f^n y, f^n x) \leq r \text{ and } d(f^n y, f^n x) \leq C\lambda^n d(y, x),\ \forall\, n \geq 0\}.\end{aligned}$$

Likewise,

$$\begin{aligned} W^u_r(x,f) &= \{y \in M \mid d(f^{-n}y, f^{-n}x) \le r,\ \forall\ n \ge 0\} \\ &= \{y \in M \mid d(f^{-n}y, f^{-n}x) \le r \text{ and } d(f^{-n}y, f^{-n}x) \le C\lambda^n d(y,x),\ \forall\ n \ge 0\}. \end{aligned}$$

Proof. We give a proof for W^s_r only. We may assume the Riemannian norm $|\cdot|$ of M is adapted to Λ. It suffices to prove there are $r > 0$, $C \ge 1$, and $0 < \lambda < 1$ such that the second set is contained in the third.

Given $x \in \Lambda$, for $y \in M$ close to x, let

$$v = exp_x^{-1}(y).$$

Then

$$d(f^n y, f^n x) = |F^n_f(v)| = |(Tf + \phi_f)^n v|,$$

where

$$\phi_f = F_f - Tf,$$

as long as the iterates make sense. It then reduces to proving there are $r > 0$, $C \ge 1$, and $0 < \lambda < 1$ such that

$$\begin{aligned} &\{v \in T_xM(r) \mid |(Tf + \phi_f)^n v| \le r,\ \forall\ n \ge 0\} \\ &\qquad \subset \{v \in T_xM(r) \mid |(Tf + \phi_f)^n v| \le C\lambda^n |v|,\ \forall\ n \ge 0\}. \end{aligned}$$

It suffices to prove there are $r > 0$, $C \ge 1$, and $0 < \lambda < 1$ such that this inclusion holds replacing the norm $|\cdot|$ by the box-adjusted norm $|\cdot|_\Lambda$ of $|\cdot|$ with respect to $T_\Lambda M = E^s \oplus E^u$.

Let $0 < \tau < 1$ be the skewness of Λ with respect to $|\cdot|$. Assume $|\cdot|_\Lambda$ is equivalent to $|\cdot|$ by a constant $K \ge 1$. Let $C = 1$, and fix

$$\tau < \lambda < 1.$$

By Lemma 4.11, there is $r > 0$ sufficiently small such that

$$\mathrm{Lip}_{2,|\cdot|}\phi_f \le K^{-2}(\lambda - \tau)$$

on $TM(Kr, |\cdot|)$. Hence

$$\mathrm{Lip}_{2,|\cdot|_\Lambda}\phi_f \le \lambda - \tau$$

on $T_\Lambda M(r, |\cdot|_\Lambda)$. (Changing the norm of a Euclidean space yields a multiplier of the relative constant for the length of vectors, but a multiplier of the square of the relative constant for the Lipschitz constants of the maps.) Then the conclusion follows directly from Lemma 4.12. This proves Theorem 4.13. □

Let X be a compact metric space. A homeomorphism $f : X \to X$ is called *expansive* if there is a constant $r > 0$ such that, for every pair of different points $x \ne y$ in X, there is an integer m such that $\mathrm{d}(f^m(x), f^m(y)) \ge r$. The number $r > 0$ is called an *expansive constant* of f.

The next theorem corresponds to Theorem 2.16.

Theorem 4.14 (Uniform expansivity of hyperbolic sets). *Let $\Lambda \subset M$ be a hyperbolic set of f. Then $f|_\Lambda$ is expansive. In fact, there are a C^1 neighborhood $\mathcal{U}_0$ of f and two numbers $a_0 > 0$ and $r_0 > 0$ such that every compact invariant set $\Delta \subset B(\Lambda, a_0)$ of every $g \in \mathcal{U}_0$ is r_0-expansive.*

Proof. We may assume the Riemannian norm $|\cdot|$ of M is adapted to Λ. By definition, $g|_\Delta$ is r_0-expansive means that if $x, y \in \Delta$ satisfy

$$d(g^n x, g^n y) \le r_0 \quad \forall n \in \mathbb{Z},$$

then $x = y$. Let $v = exp_x^{-1}(y)$. Since

$$d(g^n x, g^n y) = |F_g^n(v)| = |(Tg + \phi)^n v|,$$

where

$$\phi = \phi_g = F_g - Tg,$$

it reduces to proving there is $r_0 > 0$ such that if a vector $v \in T_x M$ satisfies

$$|(Tg + \phi)^n v| \le r_0 \quad \forall n \in \mathbb{Z},$$

then $v = 0$.

Let $0 < \tau(\Lambda) < 1$ be the skewness of Λ of Tf with respect to $|\cdot|$. Fix

$$\tau(\Lambda) < \tau_0 < \lambda < 1.$$

Let

$$\phi' = (F_g)^{-1} - (Tg)^{-1}.$$

Hence

$$(Tg)^{-1} + \phi' = (T_g + \phi)^{-1}.$$

Note that $g \to f$ implies $g^{-1} \to f^{-1}$.

By Lemma 4.8, there are a C^1 neighborhood $\mathcal{U}_0$ of f and two numbers $a_0 > 0$ and $K \ge 1$ such that any compact invariant set $\Delta \subset B(\Lambda, a_0)$ of any $g \in \mathcal{U}_0$ is hyperbolic with skewness

$$\tau(\Delta) \le \tau_0$$

with respect to $|\cdot|$, and the box-adjusted norm $|\cdot|_\Delta$ of $|\cdot|$ with respect to the hyperbolic splitting of Δ is equivalent to $|\cdot|$ with constant K. By Lemma 4.11, there are a C^1 neighborhood of f, still denoted $\mathcal{U}_0$, and a number $r_0 > 0$ such that, for any $g \in \mathcal{U}_0$,

$$\mathrm{Lip}_{2,|\cdot|}\phi \le K^{-2}(\lambda - \tau_0), \qquad \mathrm{Lip}_{2,|\cdot|}\phi' \le K^{-2}(\lambda - \tau_0)$$

on $TM(K^2 r_0; |\cdot|)$. Hence

$$\mathrm{Lip}_{2,|\cdot|_\Delta}\phi \le \lambda - \tau_0, \qquad \mathrm{Lip}_{2,|\cdot|_\Delta}\phi' \le \lambda - \tau_0$$

on $T_\Delta M(K r_0; |\cdot|_\Delta)$.

Assume a vector $v \in T_xM$, $x \in \Delta$, satisfies

$$|(Tg + \phi)^n v| \le r_0 \quad \forall n \in \mathbb{Z}.$$

Then

$$|(Tg + \phi)^n v|_\Delta \le K r_0 \quad \forall n \in \mathbb{Z}.$$

By Lemma 4.12,

$$\begin{aligned} |v|_\Delta &= |(Tg^{-1} + \phi')(T_g + \phi)(v)|_\Delta \\ &\le (\tau + \mathrm{Lip}_2\phi')|(Tg + \phi)(v)|_\Delta \le (\tau + \mathrm{Lip}_2\phi')(\tau + \mathrm{Lip}_2\phi)|v|_\Delta \le \lambda^2 |v|_\Delta. \end{aligned}$$

Here the first "$\le$" holds because, with respect to $|\cdot|_\Delta$, $(Tg+\phi)(v)$ has length $\le Kr_0$ for all negative iterates. Likewise, the second "$\le$" holds because v has length $\le Kr_0$ for all positive iterates. Since $\lambda < 1$, we get $v = 0$. This proves Theorem 4.14. □

Remark. The strong uniformness that appears in the statement of Theorem 4.14 is actually common for results about hyperbolic sets. Nevertheless for simplicity we will not state every theorem in this uniform way but only Theorems 4.14 and 4.20, just for the use of Theorem 4.23 below.

For a fiber-preserving map $\sigma : E^u \to E^s$, define the *fiber-derivative* of σ at $v \in E^u$ to be

$$D_2\sigma(v) = D(\sigma|_{E^u(x)})(v),$$

where $x = \pi v$. Also, define the *fiber-Lipschitz constant* of σ to be

$$\mathrm{Lip}_2\sigma = \sup_{x\in\Lambda} \mathrm{Lip}(\sigma|_{E^u(x)}).$$

We explained right after Lemma 4.10 that saying that $D_2\phi$ is continuous on $T_\Lambda M$ is more than saying that ϕ restricted to every fiber T_xM is C^1. Here, likewise, saying that $D_2\sigma$ is continuous on E^u is more than saying that σ restricted to every fiber $E^u(x)$ is C^1.

Now we proceed to the main part of the stable manifolds theorem for a hyperbolic set. We first consider an ideal setting $Tf + \phi : T_\Lambda M \to T_\Lambda M$, where $\phi : T_\Lambda M \to T_\Lambda M$ is continuous and fiber-preserving over f, $\phi(0_x) = 0_{fx}$, and $\mathrm{Lip}_2\phi$ is small on the whole $T_\Lambda M$. Define the (global) *fiber-unstable manifold* of $0_x \in T_xM$ to be

$$W^u(0_x, Tf + \phi) = \left\{ v \in T_xM \mid \lim_{n\to+\infty} |(Tf + \phi)^{-n} v| = 0 \right\}.$$

We remark that here the "global" fiber-unstable manifold of 0_x is on fibers, which is not the global unstable manifold $W^u(x)$ we will eventually have on the manifold M. It is just an intermediate object through which we will obtain the local unstable manifold $W^u_r(x)$ on M.

The next lemma corresponds to Lemma 2.17.

Lemma 4.15. *Let $\Lambda \subset M$ be a hyperbolic set of f with splitting $T_\Lambda M = E^u \oplus E^s$, and let $|\cdot|$ be a C^0 norm of $T_\Lambda M$ that is adapted to and of box type to $E^u \oplus E^s$. Then there is $\delta > 0$ such that:*

(1) If $\phi : T_\Lambda M \to T_\Lambda M$ is continuous, fiber-preserving over f, and fiber-Lipschitz such that

$$\mathrm{Lip}_2\phi < \delta, \quad \phi(0_x) = 0_{fx} \ \forall\, x \in \Lambda,$$

then there is a continuous fiber-preserving fiber-Lipschitz map $\sigma : E^u \to E^s$ over id, $\sigma(0_x) = 0_x$, $\mathrm{Lip}_2\sigma \le 1$, such that for any $x \in \Lambda$, $W^u(0_x, Tf + \phi)$ is exactly the graph of $\sigma_x : E^u(x) \to E^s(x)$, where $\sigma_x = \sigma|_{E^u(x)}$.

(2) If $\phi : T_\Lambda M \to T_\Lambda M$ is continuous, fiber-preserving over f, and C^1 restricted to every fiber such that

$$\mathrm{Lip}_2\phi < \delta, \quad \phi(0_x) = 0_{fx} \ \forall\, x \in \Lambda,$$

then for any $x \in \Lambda$, the map $\sigma_x = \sigma|_{E^u(x)}$ guaranteed by item (1) is C^1, and the C^1 submanifold $W^u(0_x, Tf + \phi)$ is tangent at 0_x to the unstable subspace $G^u(x)$ at x of the hyperbolic bundle isomorphism $\{T_x f + D_2\phi(0_x) \mid x \in \Lambda\}$. Moreover, if $D_2\phi$ is continuous on $T_\Lambda M$, then $D_2\sigma$ is continuous on E^u.

Proof. We first prove item (1). Let

$$\Sigma(E^u, E^s; 0) = \{\sigma : E^u \to E^s \ |\sigma \text{ is continuous, fiber-preserving over } id,\ \sigma(0_x) = 0_x,\ |\sigma|_* < \infty\},$$

where

$$|\sigma|_* = \sup_{x \in \Lambda} |\sigma_x|_*$$

and $|\sigma_x|_*$ is defined as in the proof of Lemma 2.17. With this norm $\Sigma(E^u, E^s; 0)$ forms a Banach space, and

$$\Sigma(E^u, E^s; 0)[1] = \{\sigma \in \Sigma(E^u, E^s; 0) \mid \sigma \text{ is fiber-Lipschitz, } \mathrm{Lip}_2\sigma \le 1\}$$

is a closed subset of $\Sigma(E^u, E^s; 0)$.

Let $0 < \tau < 1$ be the skewness of Λ with respect to $|\cdot|$. Let

$$\delta = \min\left\{\frac{1-\tau}{2},\ m(Tf,\ \Lambda)\right\},$$

where

$$m(Tf, \Lambda) = \inf\{m(T_x f) \mid x \in \Lambda\}.$$

This is a positive number since Λ is compact. (Later we will reduce δ further.)

Let $\phi : T_\Lambda M \to T_\Lambda M$ be continuous, fiber-preserving over f, and fiber-Lipschitz such that

$$\mathrm{Lip}_2\phi < \delta, \quad \phi(0_x) = 0_{fx} \ \forall\, x \in \Lambda.$$

We prove there is $\sigma \in \Sigma(E^u, E^s; 0)[1]$ such that its graph

$$\mathrm{gr}(\sigma) = \bigcup_{x\in\Lambda} \mathrm{gr}(\sigma_x)$$

is invariant under $Tf + \phi$; that is, for every $x \in \Lambda$,

$$(Tf + \phi)\mathrm{gr}(\sigma_x) = \mathrm{gr}(\sigma_{fx}).$$

Then we prove $\mathrm{gr}(\sigma_x)$ is exactly $W^u(0_x, Tf + \phi)$. Below, till the end of the proof of Lemma 4.15, we abbreviate

$$Tf = A.$$

The proof will match that of Lemma 2.17, like a copy.

The invariance condition

$$(A + \phi)\mathrm{gr}(\sigma_x) \subset \mathrm{gr}(\sigma_{fx})$$

is equivalent to

$$\sigma_{fx}((A + \phi)_u(v + \sigma_x v)) = (A + \phi)_s(v + \sigma_x v)$$

for every $v \in E^u(x)$. Since $A_u(\sigma_x v) = 0$ and $A_s v = 0$, this reduces to

$$\sigma_{fx}((A_{uu})_x v + \phi_u(v + \sigma_x v)) = A_{ss}(\sigma_x v) + \phi_s(v + \sigma_x v).$$

That is,

$$\sigma_{fx}((A_{uu})_x + \phi_u(I_{u,x} + \sigma_x)) = A_{ss}(\sigma_x) + \phi_s(I_{u,x} + \sigma_x).$$

Since

$$m((A_{uu})_x) \geq \tau^{-1}, \quad \mathrm{Lip}(\phi_u(I_{u,x} + \sigma_x)) \leq 2\mathrm{Lip}_2\phi < 2\delta = 1 - \tau,$$

by Theorem 2.7, $(A_{uu})_x + \phi_u(I_{u,x} + \sigma_x)$ is invertible. Hence

$$\sigma_{fx} = (A_{ss}\sigma_x + \phi_s(I_{u,x} + \sigma_x))((A_{uu})_x + \phi_u(I_{u,x} + \sigma_x))^{-1}.$$

This suggests a map

$$T = T_\phi : \Sigma(E^u, E^s; 0)[1] \to \Sigma(E^u, E^s; 0)$$

$$(T(\sigma))_{fx} = (A_{ss}\sigma_x + \phi_s(I_{u,x} + \sigma_x))((A_{uu})_x + \phi_u(I_{u,x} + \sigma_x))^{-1}, \ \forall\, x \in \Lambda,$$

called the *graph transform* induced by $A + \phi$. Finding σ with

$$(A + \phi)\mathrm{gr}(\sigma) \subset \mathrm{gr}(\sigma)$$

then reduces to finding a fixed point of T.

We verify that T maps $\Sigma(E^u, E^s; 0)[1]$ into itself. Since $\sigma_x \in \Sigma(E^u(x), E^s(x); 0_x)[1]$, it is easy to see that $(T\sigma)_{fx}(0_{fx}) = 0_{fx}$ and $(T\sigma)_{fx}$ is Lipschitz with

$$\mathrm{Lip}((T\sigma)_{fx}) \leq \frac{\tau + 2\mathrm{Lip}_2\phi}{\tau^{-1} - 2\mathrm{Lip}_2\phi} < 1.$$

Hence T maps $\Sigma(E^u, E^s; 0)[1]$ into itself.

Next we verify that T is a contraction with respect to the norm $|\cdot|_*$. For any $\sigma, \sigma' \in \Sigma(E^u, E^s; 0)[1]$, abbreviate

$$F_x = (A_{uu})_x + \phi_u(I_{u,x} + \sigma_x) : E^u(x) \to E^u(fx),$$

$$F'_x = (A_{uu})_x + \phi_u(I_{u,x} + \sigma'_x) : E^u(x) \to E^u(fx);$$

that is,

$$(T(\sigma))_{fx} F_x = A_{ss}\sigma_x + \phi_s(I_{u,x} + \sigma_x),$$

$$(T(\sigma'))_{fx} F'_x = A_{ss}\sigma'_x + \phi_s(I_{u,x} + \sigma'_x).$$

Since $F_x : E^u(x) \to E^u(fx)$ is a homeomorphism that fixes the origin, when v runs through $E^u(x) - \{0_x\}$,

$$\begin{aligned}|T(\sigma) - T(\sigma')|_* &= \sup_{x\in\Lambda}\sup_{v\neq 0_x} \frac{|(T\sigma)_x(v) - (T\sigma')_x(v)|}{|v|} \\ &= \sup_{x\in\Lambda}\sup_{v\neq 0_x} \frac{|(T\sigma)_{fx}(F_x v) - (T\sigma')_{fx}(F_x v)|}{|F_x v|}.\end{aligned}$$

On one hand,

$$\begin{aligned}&|(T\sigma)_{fx}(F_x v) - (T\sigma')_{fx}(F_x v)| \\ &\quad \le |(T\sigma)_{fx}(F_x v) - (T\sigma')_{fx}(F'_x v)| + |(T\sigma')_{fx}(F'_x v) - (T\sigma')_{fx}(F_x v)| \\ &\quad \le |A_{ss}(\sigma_x(v) - \sigma'_x(v))| + |\phi_s(v + \sigma_x(v)) - \phi_s(v + \sigma'_x(v))| \\ &\qquad + \mathrm{Lip}((T\sigma')_{fx})|F'_x v - F_x v| \\ &\quad \le \tau|\sigma_x(v) - \sigma'_x(v)| + \mathrm{Lip}\,\phi_x|\sigma_x(v) - \sigma'_x(v)| + \mathrm{Lip}\,\phi_x|\sigma_x(v) - \sigma'_x(v)| \\ &\quad \le (\tau + 2\mathrm{Lip}\,\phi_x)|\sigma_x(v) - \sigma'_x(v)|.\end{aligned}$$

On the other hand, since $\phi_x(0_x) = 0_{fx}$ and $\sigma_x(0_x) = 0_x$,

$$\begin{aligned}|F_x v| &= |(A_{uu})_x v + \phi_u(v + \sigma_x v) - \phi_u(0_x + \sigma(0_x))| \\ &\ge \tau^{-1}|v| - \mathrm{Lip}\,\phi_x(|v| + \mathrm{Lip}\,\sigma_x|v|) \\ &\ge (\tau^{-1} - 2\mathrm{Lip}\,\phi_x)|v|.\end{aligned}$$

Thus

$$\begin{aligned}|T(\sigma) - T(\sigma')|_* &\le \frac{\tau + 2\mathrm{Lip}_2\phi}{\tau^{-1} - 2\mathrm{Lip}_2\phi} \sup_{x\in\Lambda}\sup_{v\neq 0_x} \frac{|\sigma_x(v) - \sigma'_x(v)|}{|v|} \\ &= \frac{\tau + 2\mathrm{Lip}_2\phi}{\tau^{-1} - 2\mathrm{Lip}_2\phi}|\sigma - \sigma'|_*.\end{aligned}$$

Since

$$\frac{\tau + 2\mathrm{Lip}_2\phi}{\tau^{-1} - 2\mathrm{Lip}_2\phi} < 1,$$

$T = T_\phi$ is a contraction with respect to the norm $|\cdot|_*$. By the contraction mapping principle, T has a unique fixed point $\sigma = \sigma_\phi \in \Sigma(E^u, E^s; 0)[1]$ such that

$$(A + \phi)\mathrm{gr}(\sigma) \subset \mathrm{gr}(\sigma).$$

Up to this point we have done the main part of the proof of item (1), which matches Lemma 2.17, like a copy. As proved before, we actually have

$$(A + \phi)\mathrm{gr}(\sigma) = \mathrm{gr}(\sigma).$$

Moreover, reducing δ further if necessary, for every $x \in \Lambda$,

$$\mathrm{gr}(\sigma_x) = W^u(0_x, A + \phi) = W^u(0_x, Tf + \phi).$$

This proves item (1).

Before proving item (2) we insert a definition. Let $\{H_m\}_{m\in\mathbb{Z}}$ be a sequence of d-dimensional Euclidean spaces. Denote

$$H = \bigsqcup_{m\in\mathbb{Z}} H_m,$$

where $\bigsqcup$ means *discrete union*; namely, the H_m are regarded as mutually isolated so that H is a d-dimensional manifold. Let

$$A : H \to H$$

be a map such that

$$A|_{H_m} : H_m \to H_{m+1}$$

is a linear isomorphism. We call A a *hyperbolic sequence* if $\{|A_m|,\ |A_m^{-1}|\}_{m\in\mathbb{Z}}$ is bounded and, for every $m \in \mathbb{Z}$, there are a direct sum

$$H_m = E_m^s \oplus E_m^u,$$

$$A(E_m^s) = E_{m+1}^s,\ A(E_m^u) = E_{m+1}^u,$$

and two constants $C \geq 1$ and $0 < \lambda < 1$ such that

$$|A^n(v)| \leq C\lambda^n|v|, \quad \forall v \in E_m^s,\ m \in \mathbb{Z},\ n \geq 0,$$

$$|A^{-n}(v)| \leq C\lambda^n|v|, \quad \forall v \in E_m^u,\ m \in \mathbb{Z},\ n \geq 0.$$

Now we prove item (2). Assume that ϕ restricted to every fiber T_xM of $T_\Lambda M$ is C^1. We prove that σ_x is C^1 and that the C^1 submanifold $W^u(0_x, Tf+\phi)$ of T_xM is tangent at 0_x to $G^u(x)$, where $G^u(x)$ is the unstable subspace at x for the hyperbolic bundle isomorphism $\{T_xf + D_2\phi(0_x) \mid x \in \Lambda\}$. Fix $x \in \Lambda$. We put the discrete topology on $\mathrm{Orb}(x)$ and treat

$$T_{f^m x}M = H_m, \quad Tf = A.$$

In other words, we single out Tf on $T_{\mathrm{Orb}(x)}M$ to form a hyperbolic sequence. We take this point of view for the proof of item (2) because we will take derivatives of $Tf + \phi$ inside a fiber, a process that would be unnecessarily confusing if the metric of the base set is involved. The reason we can take such a point of view is that the issue of smoothness of σ_x is intrinsically inside the fiber T_xM anyway. In this setting, since the H_m are separated from each other, we may write $\mathrm{Lip}_2\phi$ or $D_2\phi$ simply as $\mathrm{Lip}\,\phi$ or $D\phi$.

Abbreviate

$$g = (Tf + \phi)|_H : H \to H.$$

Thus g is just $Tf+\phi$ expressed in the discrete setting. We call g the *discrete version* of $Tf + \phi$. By Theorem 2.7, g is a diffeomorphism of H (mapping H_m onto H_{m+1}) if $\mathrm{Lip}\,\phi$ is small. See Figure 4.5. An important observation is that g *will be Anosov if* $\mathrm{Lip}\,\phi$ *is sufficiently small.*

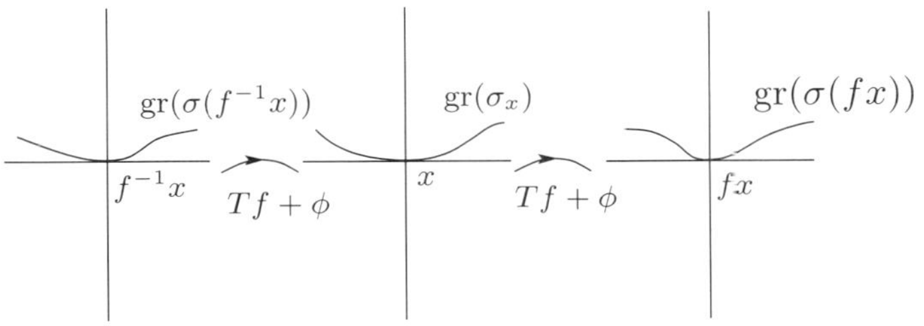

Figure 4.5. The discrete union H of linear spaces.

We verify this. The proof is the same as the proof of Lemma 2.17. For every $y \in H_m$, define

$$E^u(y) = \{y\} \times E^u(f^m x), \quad E^s(y) = \{y\} \times E^s(f^m x).$$

Then

$$T_yH = E^u(y) \oplus E^s(y), \quad y \in H.$$

This is a hyperbolic splitting for A; namely A is Anosov. If $\mathrm{Lip}\,\phi$ is sufficiently small, the same proof as in Section 4.3 will show that H is a hyperbolic set of g; namely g is Anosov. Let

$$T_yH = G^u(y) \oplus G^s(y), \quad y \in H,$$

be the hyperbolic splitting of Tg. Note that

$$\dim G^u(y) = \dim(E^u(y)).$$

Now we prove σ_x is C^1. Take any $v \in E^u(x)$. First we prove σ_x is differentiable at v. Denote $z = (v, \sigma_x(v)) \in \mathrm{gr}(\sigma_x)$ for short. By the criterion of Katok-Hasselblatt (1995) stated at the end of Chapter 2, we need to prove that the tangent set $T_z\mathrm{gr}(\sigma_x)$ is contained in a linear subspace of T_zH of dimension $\dim E^u(x)$. In fact, we prove

$$T_z\mathrm{gr}(\sigma_x) \subset G^u(z).$$

Since $\mathrm{Lip}(\sigma) \leq 1$ (here σ is the above fiber-preserving map over id; since we are in the discrete setting, we simply write $\mathrm{Lip}_2\sigma$ to be $\mathrm{Lip}\,\sigma$), for every $y \in \mathrm{gr}(\sigma)$, every generalized tangent line of $T_y\mathrm{gr}(\sigma)$ lies in the 1-cone $C_1(E^u(y))$ with respect to the direct sum $E^u(y) \oplus E^s(y)$. As long as $\mathrm{Lip}\,\phi$ is small enough, $G^u(y)$ and $G^s(y)$ will be close to $E^u(y)$ and $E^s(y)$, respectively; hence the generalized tangent lines of $T_y\mathrm{gr}(\sigma)$ will be contained in the 2-cone $C_2(G^u(y))$ with respect to the direct sum $G^u(y) \oplus G^s(y)$. Since $\mathrm{gr}(\sigma)$ is invariant under g, Tg maps generalized tangent lines of $\mathrm{gr}(\sigma)$ into generalized tangent lines of $\mathrm{gr}(\sigma)$. In particular, for every generalized tangent line l of $T_z\mathrm{gr}(\sigma_x)$,

$$Tg^{-n}(l) \subset C_2(G^u(g^{-n}z)), \ \forall n \geq 0.$$

By Theorem 4.2,

$$l \subset G^u(z).$$

This proves $T_z\mathrm{gr}(\sigma_x) \subset G^u(z)$.

By the criterion, σ_x is differentiable at v, and $\mathrm{gr}(\sigma_x)$ is tangent at z to the unstable subspace $G^u(z)$ of Tg. In particular, at the origin 0_x, $\mathrm{gr}(\sigma_x)$ is tangent to the unstable subspace $G^u(x)$ of $A + D_2\phi(0_x)$. By Theorem 4.3, $G^u(z)$ varies continuously in $z \in \mathrm{gr}(\sigma_x)$. Thus σ_x is C^1.

Finally, let $D_2\phi$ be continuous on $T_\Lambda M$. We prove $D_2\sigma$ is continuous on E^u. To this end we have to leave the discrete setting and go back to our tangent bundle $T_\Lambda M$. Note that the discrete setting is just a way of presentation. What we have proved are properties in the fibers of the tangent bundle. For instance, the direct sums $G^u(y) \oplus G^s(y)$, $y \in T_\Lambda M$, form a hyperbolic splitting for $\{T_{\pi y} f + D_2\phi(y) \mid y \in T_\Lambda M\}$. Now $D_2\phi$ is continuous on $T_\Lambda M$; hence the idea of the proof of Theorem 4.3 (hyperbolic estimates pass to the closure) shows that $G^u(y)$ varies continuously in $y \in T_\Lambda M$. Writing $z = (v, \sigma v)$ with $v \in E^u$, then $G^u(z)$ varies continuously in $v \in E^u$. Namely, $D_2\sigma(v)$ varies continuously in $v \in E^u$, proving Lemma 4.15. □

A family of embedded submanifolds $\{D_i\}_{i\in I}$ is called *self-coherent* if, for any $i, j \in I$, $\mathrm{int}\, D_i \cap \mathrm{int}\, D_j$ is open in both D_i and D_j. In particular, $\mathrm{int}\, D_i$ and $\mathrm{int}\, D_j$ do not cross each other. For instance, local solution curves of an autonomous C^1 ordinary differential equation form a family of self-coherent 1-dimensional embedded submanifolds.

The next theorem corresponds to Theorem 2.18.

Theorem 4.16 (Stable manifolds theorem for a hyperbolic set). *Let $f : M \to M$ be a C^k diffeomorphism, $k \geq 1$, and let $\Lambda \subset M$ be a hyperbolic set of f with splitting $T_\Lambda M = E^s \oplus E^u$. Then there is $r > 0$ such that, for every $x \in \Lambda$:*

(1) *$W^s_r(x)$ is a C^k embedded submanifold of M of dimension* $\dim E^s(x)$ *tangent at x to E^s_x, and $W^s_r(x)$ varies continuously in $x \in \Lambda$ with respect to the C^k topology. Precisely, there are a neighborhood V of 0_Λ in E^s and a*

continuous fiber-preserving map $\sigma : V \to E^u$ over id whose fiber-derivatives up to order k are continuous on V, and $\sigma(0_x) = 0_x$, $D_2\sigma(0_x) = 0$, such that $W_r^s(x) = exp_x \mathrm{gr}(\sigma|_{V\cap E^s(x)})$.

(2) *The family $\{W_r^s(x)\}_{x\in\Lambda}$ is self-coherent.*

(3) *The global stable manifold $W^s(x)$ is an immersed C^k submanifold of M of dimension $\dim E^s(x)$.*

Remark. Here the global stable manifold is defined by $W^s(x) = \{y \in X \mid \lim_{n\to+\infty} d(f^n y, f^n x) = 0\}$ (see the paragraph before Theorem 3.8). Theorems 4.16 and 4.13 are often combined together and referred to as the stable manifold theorem.

Let us call the map $\sigma : V \to E^u$ in item (1) the *generating map* of the family of stable manifolds $\{W_r^s(x)\}_{x\in\Lambda}$. For every $x \in \Lambda$, through the graph and the exponential map, σ C^k embeds $V \cap E^s(x)$, a neighborhood of 0_x in $E^s(x)$, right onto $W_r^s(x)$.

Proof. First we prove item (1). We first prove there is $r > 0$ such that, for every $x \in \Lambda$, $W_r^s(x)$ is a C^k submanifold of M, tangent at x to E_x^s. Since

$$W_r^s(x, f) = \exp_x(W_r^s(0_x, F_f)),$$

it suffices to prove that the fiber-stable manifold $W_r^s(0_x, F_f)$ is a C^k submanifold of T_xM, tangent at 0_x to E_x^s. We will work with $W_r^s(0_x, F_f)$ instead of $W_r^s(x, f)$ till the end of the proof of item (1).

First we prove the case $k = 1$. Let f be C^1. We first work with a norm $|\cdot|$ of $T_\Lambda M$ that is adapted to and of box type to $E^s \oplus E^u$. We extend the locally defined F_f to the whole $T_\Lambda M$ and apply item (2) of Lemma 4.15 to obtain (global) stable manifolds, then cut them off to get local stable manifolds, and finally get back to the original Riemannian norm. The proof will match that of Theorem 2.18, like a copy.

Fix a C^∞ bump function $\alpha : \mathbb{R} \to \mathbb{R}$ with $0 \le \alpha \le 1$ such that $\alpha(t) = 1$ for $|t| \le 1/3$, $\alpha(t) = 0$ for $|t| \ge 2/3$. Write

$$\phi_f = F_f - Tf : TM(r_\rho) \to TM.$$

Then ϕ_f is C^0, and C^1 restricted to fibers, and

$$\phi_f(0_x) = 0_{fx}, \quad D_2\phi_f(0_x) = 0, \quad \forall\, x \in M.$$

(Here ϕ_f has an additional property of "tangency" compared to the map ϕ of Lemma 4.15.) Define

$$\overline{\phi}_f : TM \to TM$$

$$\overline{\phi}_f(v) = \alpha\left(\frac{|v|}{3r}\right)\phi_f(v),$$

where $0 < r < r_\rho/3$ will be shrunk shortly. Then $\overline{\phi}_f$ is C^0, and C^1 restricted to fibers, and $\overline{\phi}_f = \phi_f$ on $TM(r)$. By the claim in the proof of Theorem 2.13, if $r > 0$ is sufficiently small, then $\mathrm{Lip}_2\overline{\phi}_f$ on the whole TM, with respect to $|\cdot|$, will be small enough to satisfy Lemmas 4.12 and 4.15. By Lemma 4.15, there are a continuous fiber-preserving map

$$\sigma : E^s \to E^u$$

over id, C^1 restricted to fibers, and $\sigma(0_x) = 0_x$, $\mathrm{Lip}_2\sigma \leq 1$, such that for any $x \in \Lambda$,

$$W^s(0_x, Tf + \overline{\phi}_f) = \mathrm{gr}(\sigma|_{E^s(x)}).$$

Moreover, the C^1 submanifold $W^s(0_x, Tf + \overline{\phi}_f) \subset T_xM$ is tangent at 0_x to the stable subspace of the hyperbolic linear isomorphism $Tf + D_2\overline{\phi}_f(0_x)$. Since

$$D_2\phi_f(0_x) = 0,$$

this stable subspace is just E^s_x. This means

$$D_2\sigma(0_x) = 0.$$

Since the norm $|\cdot|$ is of box type, for any $x \in \Lambda$,

$$T_xM(r) = E^s_x(r) \times E^u_x(r).$$

Denote

$$i : E^s \to T_\Lambda M$$

to be

$$i(v) = (v, \sigma(v)).$$

Then for every $x \in \Lambda$, i is a C^1 embedding that takes E^s_x onto the C^1 submanifold $\mathrm{gr}(\sigma|_{E^s(x)})$ of T_xM, tangent at 0_x to E^s_x. Since $\mathrm{Lip}\,\sigma_x \leq 1$, we have

$$i(E^s_x(r)) = W^s(0_x, Tf + \overline{\phi}_f) \cap T_xM(r).$$

We prove

$$W^s_r(0_x, F_f) = W^s(0_x, Tf + \overline{\phi}_f) \cap T_xM(r).$$

Since $\overline{\phi}_f = \phi_f$ on $T_\Lambda M(r)$, the "$\subset$" part is obvious. We prove "$\supset$". Let $v \in W^s(0_x, Tf + \overline{\phi}_f) \cap T_xM(r)$. It suffices to prove $(Tf + \overline{\phi}_f)^n v \in T_{f^n x}M(r)$ for all $n \geq 1$. But this is obvious because, by Lemma 4.12 (for the case of the whole subbundle $T_\Lambda M$),

$$W^s(0_x, Tf + \overline{\phi}_f) = \{v \in T_xM \mid |(Tf + \overline{\phi}_f)^n v| \leq (\tau + \mathrm{Lip}_2\overline{\phi}_f)^n |v|, \ \forall n \geq 0\}.$$

This proves "$\supset$". Thus

$$W^s_r(0_x, F_f) = i(E^s_x(r))$$

is a C^1 submanifold of T_xM.

So far we have worked with a norm $|\cdot|$ of $T_\Lambda M$ that is adapted to and of box type to $E^s \oplus E^u$. Now we get back to the original Riemannian norm $\|\cdot\|$ of M. Take $0 < a < r$ such that for every $x \in \Lambda$,

$$W_a^s(0_x, F_f; \|\cdot\|) \subset W_r^s(0_x, F_f; |\cdot|).$$

Then $W_a^s(0_x, F_f; \|\cdot\|)$ is a C^1 submanifold of $T_x M$. Let

$$V_x = i^{-1}(W_a^s(0_x, F_f; \|\cdot\|)).$$

Then V_x is a neighborhood of 0_x in E_x^s that is mapped onto $W_a^s(0_x, F_f; \|\cdot\|)$ by the graph of σ_x. Let

$$V = \bigcup_{x\in\Lambda} V_x.$$

Then V is a neighborhood of 0_Λ in E^s satisfying the requirement of the theorem. This proves the case $k = 1$.

Now we prove the case $k \geq 2$. Since the tangent planes have been determined in the case $k = 1$, we prove the smoothness of $W_r^s(0_x, F_f)$ only. The proof is hinted at by Robinson (1995) for a fixed point (see the proof of Theorem 2.18).

Let f be C^k. Fix $x \in \Lambda$. We prove $W_r^s(0_x, F_f)$ is a C^k submanifold of $T_x M$. Since we will take derivatives inside the fibers $T_{f^m x}M$, for clearness of presentation we take the discrete setting of the hyperbolic sequence, namely letting

$$H_m = T_{f^m x}M, \quad H = \bigsqcup_{m\in\mathbb{Z}} H_m, \quad U_m = T_{f^m x}M(r_\rho), \quad U = \bigsqcup_{m\in\mathbb{Z}} U_m, \quad 0_m = 0_{f^m x}.$$

We only consider maps $g : U \to H$ with $g : U_m \to H_{m-1}$ (fiber-preserving over the shift map $m \to m+1$). We say a C^1 map $g : U \to H$ is *tangent at the origins to a hyperbolic sequence* if

$$g(0_m) = 0_{m+1}$$

and

$$Dg(0_m) : H_m \to H_{m+1}$$

is a hyperbolic sequence. Since the discrete version of F_f is such a map, it suffices to prove inductively that *if $g : U \to H$ is C^k and is tangent at the origins to a hyperbolic sequence, then there is $r > 0$ such that $W_r^s(0_0, g)$ is a C^k submanifold of H_0*. Note that $0_0 = 0_x$.

The case $k = 1$ is just proved (by applying the result of Lemma 4.15, hence without using the discrete setting of the hyperbolic sequence). Assume that the case of $k - 1$ is proved; we prove the case of k. Let g be C^k. The proof will be like a copy of that of Theorem 2.18.

Define a map G such that, for every $m \in \mathbb{Z}$,

$$G : U_m \times H_m \to H_{m+1} \times H_{m+1}$$
$$G(y, v) = (g(y),\ Dg(y)v).$$

Here the product spaces take the usual max metric. Then G is C^{k-1}. Clearly,

$$G^n(y, v) = (g^n(y),\ Dg^n(y)v).$$

Also,

$$G(0_m, 0_m) = (0_{m+1}, 0_{m+1}),$$

and

$$DG(0_m, 0_m) = \begin{pmatrix} Dg(0_m) & 0 \\ 0 & Dg(0_m) \end{pmatrix}.$$

Hence $\{DG(0_m, 0_m)\}_{m\in\mathbb{Z}}$ is a hyperbolic sequence. By induction, there is $r > 0$ such that $W_r^s((0_0, 0_0), G)$ is a C^{k-1} submanifold of $U_0 \times H_0$.

On the other hand, by Theorem 4.13, we may assume $r > 0$ has been chosen such that

$$W_r^s((0_0, 0_0),\ G) = \{(y, v) \in U_0 \times H_0 \mid |g^n y| \le r,\ |Dg^n(y)v| \le r,\ \forall n \ge 0\},$$

where $Dg^n(y)v$ is just $Tg^n(v)$. As proved before, g always agrees with an Anosov diffeomorphism $\overline{g}$ of H on a neighborhood of the set of origins 0_m. Let

$$T_yH = G^u(y) \oplus G^s(y), \quad y \in H,$$

be the hyperbolic splitting of $\overline{g}$. If $r > 0$ is sufficiently small, then $v \in T_yH$ satisfies

$$|Tg^n(v)| \le r,\ \forall n \ge 0$$

if and only if

$$v \in G^s(y)(r).$$

Hence the above equation can be rewritten as

$$W_r^s((0_0, 0_0), G) = \{(y, v) \in U_0 \times H_0 \mid y \in W_r^s(0_0, g),\ v \in G^s(y)(r)\}.$$

As proved before,

$$G^s(y) = T_y(W^s(0_0, g)).$$

Hence $W_r^s((0_0, 0_0), G)$ is just (the r-neighborhood of the 0-set of) the tangent bundle of $W_r^s(0_0, g)$. Since the degrees of smoothness of a manifold and its tangent bundle differ by 1 and since $W_r^s((0_0, 0_0), G)$ is C^{k-1}, so $W_r^s(0_0, g)$ is C^k. This proves the case of k.

So far we have proved that if f is C^k, then $W_r^s(0_x, F_f)$ is a C^k submanifold of T_xM. There is one more step to go, that is, to prove that the fiber-derivatives of σ of Theorem 4.16 up to order k are continuous on $E^s(r)$.

Since the problem restricted to fibers is just solved, it remains to prove that these fiber-derivatives are continuous with respect to the base points $x \in \Lambda$. To this end we have to leave the discrete setting and go back to our tangent bundle. But before that let us use the discrete setting to define a map Σ. Let g and G be as above. For simplicity we use the same notation σ to denote its discrete version. Thus $\sigma : E^s_m(r) \to E^u_m(r)$ is the map with $\mathrm{gr}(\sigma) = W^s_r(0_m, g)$, where $E^s_m \oplus E^u_m$ is the hyperbolic splitting of $Dg(0_m)$ and $r > 0$ is the small number just determined. Define

$$\Sigma : E^s_m(r) \times E^s_m(r) \to E^u_m \times E^u_m$$
$$\Sigma(a, b) = (\sigma(a),\ D\sigma(a)b).$$

Claim. Σ is the *generating map* of G, meaning that for every $m \in \mathbb{Z}$, $\mathrm{gr}(\Sigma_m) = W^s_r((0_m, 0_m), G)$, where $\Sigma_m = \Sigma|_{E^s_m(r)\times E^s_m(r)}$.

In fact, by definition, $(y, v) \in \mathrm{gr}(\Sigma_m)$ if and only if

$$y^u = \sigma(y^s), \quad y^s \in E^s_m(r) \ \text{ and } \ v^u = D\sigma(y^s)v^s, \quad v^s \in E^s_m(r).$$

That is,

$$y \in \mathrm{gr}(\sigma) = W^s_r(0_m, g) \ \text{ and } \ v \in \mathrm{gr}(D\sigma(y^s)|_{E^s_m(r)}) = G^s(y)(r).$$

This means

$$(y, v) \in W^s_r((0_m, 0_m), G),$$

proving the claim.

While a fiber-preserving map on the bundle has its discrete version, conversely, the map G defined in the discrete setting gives a map

$$G_\Lambda : T_xM(\rho) \times T_xM \to T_{fx}M \times T_{fx}M, \ \forall\, x \in \Lambda,$$

such that G is the discrete version of G_Λ. Likewise, the map Σ gives a map Σ_Λ in the bundle such that Σ is the discrete version of Σ_Λ.

Now we go back to our tangent bundle $T_\Lambda M$. Let f be C^k. We prove that the fiber-derivatives of σ up to order k are continuous in $x \in \Lambda$. We sketch the proof. Since F_f has fiber-derivatives up to order k continuous in $x \in \Lambda$ and since its discrete version is tangent at the origins to a uniform family of hyperbolic sequences, it suffices to prove inductively that *if a fiber-preserving map F of a bundle over $f : \Lambda \to \Lambda$ defined near the origins has fiber-derivatives up to order k continuous in $x \in \Lambda$ and if the discrete version of F is tangent at the origins to a uniform family of hyperbolic sequences, then there is $r > 0$ such that the generating map σ of F of size r has fiber-derivatives up to order k continuous in $x \in \Lambda$.* Note that we have not specified the bundle for F because the induction will concern a collection of bundles over the same base Λ, such as the one where the map G_Λ lives.

The case $k = 1$ is guaranteed by item (2) of Lemma 4.15. Assume that the case of $k - 1$ is proved. We prove the case of k. Thus, assume that F has fiber-derivatives up to order k continuous in $x \in \Lambda$ and that its discrete version g is tangent at the origins to a uniform family of hyperbolic sequences. Let G be the map induced by g as above. Since G is defined using g and Dg, which are just F and D_2F in the bundle, the fiber-derivatives of G_Λ up to order $k-1$ are made up of fiber-derivatives of F up to order k and hence are continuous in $x \in \Lambda$. As proved above, G is tangent at the origins to a uniform family of hyperbolic sequences and, by the claim, Σ_Λ is the generating map of G_Λ. Thus, by induction, there is $r > 0$ such that the generating map Σ_Λ of size r has fiber-derivatives up to order $k-1$ continuous in $x \in \Lambda$. By the definition of Σ_Λ, σ has fiber-derivatives up to order k continuous in $x \in \Lambda$. This proves item (1) of Theorem 4.16.

Now we prove item (3). It is easy to see that, for any $r > 0$,

$$W^s(x) = \bigcup_{n\geq 0} f^{-n}W^s_r(f^n x), \quad W^u(x) = \bigcup_{n\geq 0} f^n W^u_r(f^{-n}x).$$

Now Λ is a hyperbolic set and $x \in \Lambda$. By item (1), $W^s_r(x)$ is a C^k embedded submanifold of M. Hence $f^{-n}W^s_r(x)$ is a C^k embedded submanifold for every $n \geq 0$. As a monotone union of a sequence of embedded submanifolds, $W^s(x)$ is a C^k immersed submanifold of M. Likewise for $W^u(x)$.

Finally we prove item (2); that is, the family of embedded submanifolds $\{W^s_r(x)\}_{x\in\Lambda}$ is self-coherent. Note that while $W^s_r(0_x) \subset T_xM$ and $W^s_r(0_y) \subset T_yM$ are by definition disjoint for $x \neq y$, $W^s_r(x) = exp_x(W^s_r(0_x))$ and $W^s_r(y) = exp_y(W^s_r(0_y))$ in M do intersect if $d(x,y)$ is small. Thus there is indeed a problem of coherence of intersections.

Let $W^s_r(x) \cap W^s_r(y) \neq \emptyset$, where $x, y \in \Lambda$. Then both $W^s_r(x)$ and $W^s_r(y)$ are contained in the global stable manifold $W^s(x)$. Since the three submanifolds have the same dimension, the coherence of the family $W^s_r(x)$, $x \in \Lambda$, follows immediately. This proves Theorem 4.16. □

Remark. The proof of the stable manifolds theorem for a hyperbolic set is long, going through Lemma 4.15 and Theorem 4.16. It follows the four steps of the proof for a hyperbolic fixed point, summarized after the proof of Theorem 2.18. There are some additional issues involved here, such as transferring the map f on the manifold to the self-lifting F_f on the tangent bundle, establishing a discrete setting to investigate the smoothness of the fiber-stable manifolds $W^s_r(0_x, F_f)$, comparing the tangent bundle and the discrete setting to investigate the continuous dependence on base points x for the fiber-derivatives of the generating map of the family $\{W^s_r(0_x, F_f)\}_{x\in\Lambda}$, exploring the self-coherence of the family $\{W^s_r(x)\}_{x\in\Lambda}$, etc. Nevertheless the main part of the proof still matches the proof for a hyperbolic fixed point, like a copy.

Theorem 4.17. *Let $\Lambda \subset M$ be a hyperbolic set of f. There are $r > 0$ and $\delta > 0$ such that for any $x, y \in \Lambda$, if $d(x, y) \leq \delta$, then $W_r^s(x) \cap W_r^u(y) \neq \emptyset$ transversely.*

Proof. Let $\Lambda_i = \{x \in \Lambda : \dim E^s(x) = i\}$. Then $\Lambda_0, \ldots, \Lambda_{\dim M}$ are finitely many disjoint compact invariant sets. By Theorem 4.1, Λ_0 and $\Lambda_{\dim M}$ are of finite many points. Thus we may assume all points of Λ have the same index $1 \leq i \leq \dim M - 1$.

By Theorem 4.16, there is $r > 0$ such that, for every $x \in \Lambda$, $W_r^s(x)$ and $W_r^u(x)$ are C^1 submanifolds of M and vary continuously in x with respect to the C^1 topology. Since $W_r^s(x)$ and $W_r^u(x)$ intersect transversely at x, there is $\delta(x) > 0$ such that if $y \in \Lambda$ satisfies $d(x, y) \leq \delta(x)$, then $W_r^s(x) \cap W_r^u(y) \neq \emptyset$ transversely. Since Λ is compact, $\delta(x)$ can be chosen independent of $x \in \Lambda$. This proves Theorem 4.17. □

There is a nice geometrical explanation for the expansiveness of a hyperbolic set using local stable manifolds. Let Λ be a hyperbolic set of f. Since $W_r^s(x)$ and $W_r^u(x)$ intersect transversely at x, if $r > 0$ is small enough, then $W_r^s(x) \cap W_r^u(x) = \{x\}$ for any $x \in \Lambda$. If $d(f^n y, f^n x) \leq r$ for all $n \in \mathbb{Z}$, then $y \in W_r^s(x) \cap W_r^u(x)$ by Theorem 4.13. Hence $y = x$ proving that f is r-expansive on Λ.

The best picture for the family of global stable manifolds $\{W^s(x, f)\}$ is presented when f is Anosov. In that case Theorem 4.16 applies to every point and hence yields the local stable manifold $W_r^s(x)$ for every $x \in M$. The self-coherence of $\{W_r^s(x)\}$ then guarantees that the family of global stable manifolds $W^s(x)$ forms a C^0 foliation of the whole manifold M with C^k leaves, a beautiful global picture.

(By definition, for $m < n$, a decomposition of an n-dimensional manifold M into a disjoint union of m-dimensional C^k immersed submanifolds of M is called an m-dimensional C^0 *foliation* with C^k *leaves* if, in C^0 local charts, the union is like $\mathbb{R}^n = \bigcup(\mathbb{R}^m \times \{c\})$, $c \in \mathbb{R}^{n-m}$.)

4.5. Structural stability of hyperbolic sets

First we consider sections of the tangent bundle.

Let $\Lambda \subset M$ be a compact invariant set of f. As usual, by a *section* (or *vector field*) of $T_\Lambda M$ we mean a map $\gamma : \Lambda \to T_\Lambda M$ such that $\gamma(x) \in T_x M$ for every $x \in \Lambda$. Denote by $\Gamma^0(T_\Lambda M)$ the Banach space of all continuous sections of $T_\Lambda M$, endowed with the C^0 norm

$$|\gamma| = \sup_{x \in \Lambda} |\gamma(x)|.$$

Let $F : T_\Lambda M \to T_\Lambda M$ be a fiber-preserving map that covers $f : \Lambda \to \Lambda$. We call $\gamma \in \Gamma^0(T_\Lambda M)$ an *invariant section* of F if

$$F(\gamma(x)) = \gamma(fx), \ \forall \ x \in \Lambda.$$

Omitting base points, this can be written

$$F(\gamma) = \gamma.$$

For instance, the zero section is invariant under Tf.

Recall our convention at the beginning of Section 4.4 that hyperbolic sets in M are always assumed to be compact. Also recall that $T_\Lambda M(r)$ denotes the set of vectors of $T_\Lambda M$ with norm $\leq r$.

The next lemma corresponds to Lemma 2.5.

Lemma 4.18. *Let $\Lambda \subset M$ be a hyperbolic set of f with splitting $T_\Lambda M = E^s \oplus E^u$, and let $|\cdot|$ be a C^0 norm of $T_\Lambda M$ that is adapted to and of box type to $E^s \oplus E^u$. Let $0 < \tau < 1$ be the skewness of Λ with respect to $|\cdot|$. Let $r > 0$. If $\phi : T_\Lambda M(r) \to T_\Lambda M$ is continuous, fiber-preserving over f, and fiber-Lipschitz with*

$$\mathrm{Lip}_2\phi < 1 - \tau,$$

then $Tf + \phi$ has in $\Gamma^0(T_\Lambda M)(r)$ at most one invariant section. If, in addition,

$$|\phi(0_\Lambda)| = \sup_{x\in\Lambda}|\phi(0_x)| \leq (1 - \tau - \mathrm{Lip}_2\phi)r,$$

then $Tf + \phi$ has in $\Gamma^0(T_\Lambda M)(r)$ at least one (hence a unique) invariant section γ_ϕ. Moreover,

$$|\gamma_\phi| \leq \frac{|\phi(0_\Lambda)|}{1 - \tau - \mathrm{Lip}_2\phi}.$$

Proof. The proof will be like a copy of that of Lemma 2.5. We solve the equation

$$(Tf + \phi)\gamma = \gamma$$

for $\gamma \in \Gamma^0(T_\Lambda M)(r)$. Till the end of the proof of Lemma 4.18, we abbreviate

$$Tf = A.$$

It is equivalent to solve for γ such that

$$(A + \phi)\gamma(x) = \gamma(fx), \ \forall x \in \Lambda.$$

Writing in components with respect to the direct sum

$$T_\Lambda M = E^s \oplus E^u,$$

this is equivalent to solving

$$A_s\gamma(x) + \phi_s\gamma(x) = \gamma_s(fx), \ A_u\gamma(x) + \phi_u\gamma(x) = \gamma_u(fx), \ \forall x \in \Lambda,$$

or

$$A_{ss}\gamma_s(x) + \phi_s\gamma(x) = \gamma_s(fx), \ A_{uu}\gamma_u(x) + \phi_u\gamma(x) = \gamma_u(fx), \ \forall x \in \Lambda,$$

or

$$A_{ss}\gamma_s(x) + \phi_s\gamma(x) = \gamma_s(fx),\ A_{uu}^{-1}\gamma_u(fx) - A_{uu}^{-1}\phi_u\gamma(x) = \gamma_u(x),\ \forall x \in \Lambda.$$

Of course the first equation can be also written as

$$A_{ss}\gamma_s(f^{-1}x) + \phi_s\gamma(f^{-1}x) = \gamma_s(x),$$

in order to "match" the base points at the right-hand side of the second equation.

This suggests a map

$$T = T_\phi : \Gamma^0(T_\Lambda M)(r) \to \Gamma^0(T_\Lambda M)$$

$$T(\gamma)(x) = (A_{ss}\gamma_s(f^{-1}x) + \phi_s\gamma(f^{-1}x),\ A_{uu}^{-1}\gamma_u(fx) - A_{uu}^{-1}\phi_u\gamma(x)),\ \forall x \in \Lambda.$$

Then T and $A+\phi$ have the same set of invariant sections. Therefore, to prove that $A+\phi$ has at most one invariant section in $\Gamma^0(T_\Lambda M)(r)$, it suffices to prove T is a contraction. Since

$$|T(\gamma) - T(\gamma')| = \sup_{x\in\Lambda}|T(\gamma)(x) - T(\gamma')(x)|,$$

where the E^s-part is less than or equal to

$$\sup_{x\in\Lambda}\ (\tau\ \gamma_s(f^{-1}x) - \gamma_s'(f^{-1}x)| + \mathrm{Lip}_2\phi \cdot |\gamma(f^{-1}x) - \gamma'(f^{-1}x)|)$$

and the E^u-part is less than or equal to

$$\sup_{x\in\Lambda}\ (\tau|\gamma_u(fx) - \gamma_u'(fx)| + \tau\mathrm{Lip}_2\phi \cdot |\gamma(x) - \gamma'(x)|);$$

hence

$$|T(\gamma) - T(\gamma')| \le (\tau + \mathrm{Lip}_2\phi)|\gamma - \gamma'|.$$

Since

$$\mathrm{Lip}_2\phi < 1 - \tau,$$

T is indeed a contraction. This proves that $A+\phi$ has in $\Gamma^0(T_\Lambda M)(r)$ at most one invariant section.

Now assume in addition that

$$|\phi(0_\Lambda)| \le (1 - \tau - \mathrm{Lip}_2\phi)r.$$

We prove T maps $\Gamma^0(T_\Lambda M)(r)$ into itself. Take any $\gamma \in \Gamma^0(T_\Lambda M)(r)$. Since, for any $x \in \Lambda$,

$$|T(0_\Lambda)(x)| = |(\phi_s(0_{f^{-1}x}),\ -A_{uu}^{-1}\phi_u(0_x))| \le |\phi(0_\Lambda)|,$$

we have

$$\begin{aligned} |T(\gamma)| &\le |T(0_\Lambda)| + |T(\gamma) - T(0_\Lambda)| \\ &\le |\phi(0_\Lambda)| + (\tau + \mathrm{Lip}_2\phi)|\gamma| \le r. \end{aligned}$$

Thus T maps $\Gamma^0(T_\Lambda M)(r)$ into itself. By the contraction mapping principle, T has a (unique) fixed point γ_ϕ in $\Gamma^0(T_\Lambda M)(r)$. Moreover, letting $\gamma = \gamma_\phi$ in the above inequality

$$|T(\gamma)| \leq |\phi(0_\Lambda)| + (\tau + \mathrm{Lip}_2\phi)|\gamma|$$

gives

$$|\gamma_\phi| \leq |\phi(0_\Lambda)| + (\tau + \mathrm{Lip}_2\phi)|\gamma_\phi|.$$

Hence

$$|\gamma_\phi| \leq \frac{|\phi(0_\Lambda)|}{1 - \tau - \mathrm{Lip}_2\phi}.$$

This proves Lemma 4.18. □

Now we study structural stability of hyperbolic sets.

Let $\Lambda \subset M$ be a compact invariant set of f. We say f is C^r *embeddingly stable* on Λ if there is a C^r neighborhood $\mathcal{U}$ of f such that for any $g \in \mathcal{U}$, there is a continuous injective map $h = h_g : \Lambda \to M$ such that $hf = gh$ on Λ. In this case, $h(\Lambda)$ will be a compact invariant set of g. Since Λ is compact, $h : \Lambda \to h(\Lambda)$ is a homeomorphism. That is, h is a topological conjugacy from $f|_\Lambda$ to $g|_{h(\Lambda)}$. Since the set $h(\Lambda)$ comes as the image of h, but not specified beforehand, we have called such a stability an embedding one.

Formally stronger is the notion of ϵ-embedding stability. We say f is C^r *ϵ-embeddingly stable* on Λ if, for any $\epsilon > 0$, there is a C^r neighborhood $\mathcal{U}$ of f such that for any $g \in \mathcal{U}$, there is a continuous injective map $h = h_g : \Lambda \to M$ such that $hf = gh$ and $d(h, id) \leq \epsilon$.

A fundamental result for the structural stability theory is that every hyperbolic set is C^1 ϵ-embeddingly stable. More precisely,

Theorem 4.19 (The embedding stability of hyperbolic sets). *Let $\Lambda \subset M$ be a hyperbolic set of f. There are a C^1 neighborhood $\mathcal{U}_0$ of f and a number $\epsilon_0 > 0$ such that, for every $g \in \mathcal{U}_0$, there is a unique continuous injective map $h = h_g : \Lambda \to M$ that satisfies $hf = gh$ and $d(h, id) \leq \epsilon_0$. Moreover, $d(h_g, id) \to 0$ when $d^1(g, f) \to 0$.*

The image $h_g(\Lambda)$ is a compact invariant set of g, called the *continuation* of Λ under g, often denoted by Λ_g. Thus Λ_g is defined for g sufficiently C^1 close to f.

A direct consequence of Theorem 4.19 is the celebrated theorem of Anosov (1967) stated next. Recall from topology the theorem of *invariance of domain* stating that if $U \subset \mathbb{R}^n$ is an open subset and $f : U \to \mathbb{R}^n$ is continuous and injective, then $f(U)$ is an open subset of $\mathbb{R}^n$ and the inverse of f is continuous. It can be found in many textbooks on algebraic topology. See for instance Munkres (1984).

Theorem 4.20. *Anosov diffeomorphisms are C^1 structurally stable.*

Proof. Let $f: M \to M$ be Anosov. By Theorem 4.19, there is a C^1 neighborhood $\mathcal{U}$ of f such that for any $g \in \mathcal{U}$, there is a continuous injective map $h = h_g : M \to M$ such that $hf = gh$. We only need prove h is onto. By the invariance of domain, h is an open map. Hence $h(M)$ is both closed and open. Therefore $h(M) = M$ (we assume M is connected). □

Many authors made important contributions to Theorem 4.19. Anosov (1967) proved the theorem for the case when the hyperbolic set is the whole manifold, using a geometrical approach. Moser (1969) came up with a functional analysis approach for Anosov's result, reducing the problem to an ingenious application of the contraction mapping principle of an infinite dimensional Banach space. Subsequent extensions of Moser's approach were given by Mather (1967), Hirsch-Pugh (1970), Robbin (1971), and perhaps others. Bowen (1975) carried over the geometrical idea of Anosov to a general hyperbolic set.

Let us explain the ideas of Moser, et al. Given a C^1 perturbation g of f, we are to solve for h from the conjugacy equation

$$gh = hf,$$

where h belongs to the set $C^0(\Lambda, M)$ of continuous maps from Λ to M, which is not a linear space and hence hard to work with. However, a map $h \in C^0(\Lambda, M)$ near the identity id_Λ gives rise to, through the exponential map, a section $\gamma \in \Gamma^0(T_\Lambda M)$ near the zero-section 0_Λ, and vice versa:

$$\gamma(x) = exp_x^{-1} h(x), \quad h(x) = exp_x \gamma(x), \ \forall\, x \in \Delta.$$

Thus the problem reduces to solving the equation

$$g\ exp_x \gamma(x) = exp_{fx} \gamma(fx),$$

or

$$(*) \qquad \underbrace{exp_{fx}^{-1}\ g\ exp_x}_{F_f^g} \gamma(x) = \gamma(fx), \ \forall\, x \in \Lambda$$

for γ, which lives in a small ball about the origin of $\Gamma^0(T_\Lambda M)$. This is to solve for an invariant section of the fiber-preserving map F_f^g in $(*)$.

Here is the precise definition of F_f^g. Let $f: M \to M$ and $g: M \to M$ be two diffeomorphisms that are close to each other, and let $r > 0$ be small such that, for any $x, y \in M$, if $d(x, y) \le r$, then $d(fx, gy) \le \rho$. Here $\rho > 0$ is the constant in Theorem 4.9. Define the *lifting*

$$F_f^g : TM(r) \to TM$$

of g over f to be

$$F_f^g(v) = exp_{f(\pi v)}^{-1}\ g\ exp_{\pi v}(v),$$

where $\pi : TM \to M$ is the bundle projection. Clearly F_f^g is C^1 and fiber-preserving over f. Briefly, treating orbits of f as "moving origins", F_f^g exhibits the behavior of g near these origins.

Now, to solve for the invariant section of F_f^g in $(*)$, by Lemma 4.18, it suffices to show that, near the 0-section, F_f^g is a fiber-Lipschitz perturbation of Tf, which is not surprising because if $g = f$, then F_f^f is just F_f.

Before proving Theorem 4.19 note that, while Theorem 4.19 has a neat statement and many important applications like Theorem 4.20, it is not sufficient to guarantee the structural stability theorem of isolated hyperbolic sets (Theorem 4.23). Indeed, to prove Theorem 4.23 we will need to apply Theorem 4.19 to the continuation Λ_g of g and hence to get a C^1 neighborhood $\mathcal{U}_1$ of g. However, it is not clear whether $\mathcal{U}_1$ has sufficient size so that $f \in \mathcal{U}_1$; hence it is not clear whether there is an embedding conjugacy taking Λ_g back into an invariant set of f.

For this reason, we will not prove Theorem 4.19 but a stronger result instead:

Theorem 4.21. *Let Λ be a hyperbolic set of f. There are a C^1 neighborhood $\mathcal{U}_0$ of f and two numbers $a_0 > 0$ and $\epsilon_0 > 0$ such that for any $g, g' \in \mathcal{U}_0$ and any compact invariant set $\Delta \subset B(\Lambda, a_0)$ of g, there is at most one continuous map $h : \Delta \to M$ such that $hg = g'h$ and $d(h, id) \le \epsilon_0$. Moreover, for any $0 < \epsilon \le \epsilon_0$, there is a C^1 neighborhood $\mathcal{U} \subset \mathcal{U}_0$ of f such that for any $g, g' \in \mathcal{U}$ and any compact invariant set $\Delta \subset B(\Lambda, a_0)$ of g, there is at least one (hence unique) continuous injective map $h : \Delta \to M$ such that $hg = g'h$ and $d(h, id) \le \epsilon$.*

This amounts to a uniform version of Theorem 4.19 with respect to perturbations, which automatically implies Theorem 4.19. It is in the spirit of the highly uniform statement of the (big) shadowing theorem of Anosov (1970) (see Katok-Hasselblatt (1995), page 566), while adopting the form of the statement of Theorem 2.6.

Proof. Though the statement of Theorem 4.21 looks much more complicated than Theorem 4.19, the proof will be along the same idea as just explained.

We may assume that the Riemannian norm $|\cdot|$ of M is adapted to Λ. First take $\mathcal{U}_0$ and r_0 sufficiently small such that, for any $g, g' \in \mathcal{U}_0$, the lifting

$$F_g^{g'} : TM(r_0) \to TM$$

$$F_g^{g'}(v) = exp_{g(\pi v)}^{-1}\ g'\ exp_{\pi v}(v)$$

can be defined. Let

$$\phi = \phi_{g,g'} = F_g^{g'} - Tg : TM(r_0) \to TM.$$

Then ϕ is fiber-preserving over g. For simplicity we omit the subscript g, g' of ϕ.

Claim. *For any $\epsilon > 0$, there are a C^1 neighborhood $\mathcal{U}$ of f and a number $r > 0$ such that, for any $g, g' \in \mathcal{U}$, $\mathrm{Lip}_2\phi < \epsilon$ on $TM(r)$.*

In fact, by the proof of Lemma 4.11, there are a C^1 neighborhood $\mathcal{U}$ of f and a number $r > 0$ such that for any $g \in \mathcal{U}$,

$$|D_2(F_g)(v) - Tg(v)| < \epsilon/2$$

on $TM(r)$. Now

$$D_2(F_g)(v) = D(exp_{gx}^{-1})(g(exp_x v)) \circ Dg(exp_x v) \circ D(exp_x)(v),$$

$$D_2(F_g^{g'})(v) = D(exp_{gx}^{-1})(g'(exp_x v)) \circ Dg'(exp_x v) \circ D(exp_x)(v).$$

Hence we may assume $\mathcal{U}$ has been chosen sufficiently small so that if $g, g' \in \mathcal{U}$, then

$$|D_2(F_g^{g'})(v) - D_2(F_g)(v)| < \epsilon/2$$

on $TM(r)$. Hence

$$\begin{aligned} |D_2\phi(v)| &= |D_2(F_g^{g'})(v) - Tg(v)| \\ &\le |D_2(F_g^{g'})(v) - D_2(F_g)(v)| + |D_2(F_g)(v) - Tg(v)| < \epsilon. \end{aligned}$$

Applying the generalized mean value theorem to fibers, we get $\mathrm{Lip}_2\phi < \epsilon$ on $TM(r)$. This proves the claim.

Let $0 < \tau(\Lambda) < 1$ be the skewness of Λ under f with respect to $|\cdot|$. Fix

$$\tau(\Lambda) < \tau_0 < \lambda < 1.$$

By Lemma 4.8, there are a C^1 neighborhood $\mathcal{U}_0$ of f and two numbers $a_0 > 0$ and $K \ge 1$ such that any compact invariant set $\Delta \subset B(\Lambda, a_0)$ of any $g \in \mathcal{U}_0$ is hyperbolic with skewness

$$\tau(\Delta) \le \tau_0$$

with respect to $|\cdot|$, and the box-adjusted norm $|\cdot|_\Delta$ of $|\cdot|$ with respect to the hyperbolic splitting of Δ is equivalent to $|\cdot|$ with constant K. By the claim, there are a C^1 neighborhood of f, still denoted $\mathcal{U}_0$, and a number $r_0 > 0$ such that for any $g, g' \in \mathcal{U}_0$,

$$\mathrm{Lip}_{2,|\cdot|}\phi \le K^{-2}(\lambda - \tau_0),$$

on $TM(K^2 r_0; |\cdot|)$, where $\phi = \phi_{g,g'} = F_g^{g'} - Tg$. Hence

$$\mathrm{Lip}_{2,|\cdot|_\Delta}\phi \le \lambda - \tau_0$$

on $T_\Delta M(Kr_0; |\cdot|_\Delta)$. By Theorem 4.14, we may assume $\mathcal{U}_0$, a_0, and r_0 have been chosen sufficiently small so that any compact invariant set $\Delta \subset B(\Lambda, a_0)$ of any $g \in \mathcal{U}_0$ is r_0-expansive. Then let

$$\epsilon_0 = r_0/2.$$

Up to this point $\mathcal{U}_0$, a_0, and ϵ_0 are settled.

Now for any $g, g' \in \mathcal{U}_0$ and any compact invariant set $\Delta \subset B(\Lambda, a_0)$ of g,

$$\mathrm{Lip}_{2,|\cdot|_\Delta}\phi \le \lambda - \tau_0 \le \lambda - \tau(\Delta)$$

on $T_\Delta M(K\epsilon_0; |\cdot|_\Delta)$. Hence by Lemma 4.18, $F_g^{g'} = Tg + \phi$ has at most one invariant section in $\Gamma^0(T_\Delta M)(K\epsilon_0; |\cdot|_\Delta)$, hence at most one invariant section in $\Gamma^0(T_\Delta M)(\epsilon_0; |\cdot|)$, denoted γ. That is,

$$exp_{gx}^{-1}\, g'\, exp_x\gamma(x) = \gamma(gx), \ \forall\ x \in \Delta.$$

Let $h(x) = exp_x\gamma(x)$. This means there is at most one continuous map $h : \Delta \to M$ such that $d(h, id) \le \epsilon_0$ and

$$g'h(x) = hg(x), \ \forall\ x \in \Delta.$$

Let $0 < \epsilon \le \epsilon_0$ be given. Take a C^1 neighborhood $\mathcal{U} \subset \mathcal{U}_0$ of f such that, for every $g \in \mathcal{U}$,

$$\sup_{x\in M} d(fx,\ gx) < \frac{(1-\lambda)\epsilon}{2K^2}.$$

For any $g, g' \in \mathcal{U}$ and any compact invariant set $\Delta \subset B(\Lambda, a_0)$ of g,

$$\begin{aligned}|\phi(0_\Delta)|_\Delta &= \sup\nolimits_{x\in\Delta}|\phi(0_x)|_\Delta = \sup_{x\in\Delta}|F_g^{g'}(0_x)|_\Delta \\ &= \sup_{x\in\Delta}|exp_{gx}^{-1}(g'x)|_\Delta \le K \sup_{x\in\Delta}|exp_{gx}^{-1}(g'x)| \\ &= K\sup_{x\in\Delta} d(gx, g'x) \le (1-\lambda)K^{-1}\epsilon.\end{aligned}$$

By Lemma 4.18, $F_g^{g'}$ has in $\Gamma^0(T_\Delta M)(K\epsilon_0, |\cdot|_\Delta)$ at least one (hence a unique) invariant section γ, and

$$|\gamma|_\Delta \le \frac{(1-\lambda)K^{-1}\epsilon}{1-\tau(\Delta)-\mathrm{Lip}_{2,|\cdot|_\Delta}\phi} \le K^{-1}\epsilon.$$

Hence

$$|\gamma| \le \epsilon.$$

This gives a continuous map $h : \Delta \to M$ such that $d(h, id) \le \epsilon$ and

$$g'h(x) = hg(x), \ \forall\ x \in \Delta.$$

We prove h is injective. In fact, assume there are $x,\ y \in \Delta$ with $h(x) = h(y)$. Then for any $n \in \mathbb{Z}$,

$$\begin{aligned} d(g^n x,\ g^n y) &\leq d(g^n x,\ h(g^n x)) + d(h(g^n x),\ h(g^n y)) + d(h(g^n y),\ g^n y) \\ &= d(g^n x,\ h(g^n x)) + d(g'^n(hx),\ g'^n(hy)) + d(h(g^n y),\ g^n y) \\ &\leq \epsilon + 0 + \epsilon \leq 2\epsilon_0 = r_0. \end{aligned}$$

Since $g|_\Delta$ is r_0-expansive, we get $x = y$, proving h is injective. This proves Theorem 4.21. □

The proof of Theorem 4.21 looks complicated since it involves two norms. Nevertheless the core, the proof of Lemma 4.18, is still a copy of that of Lemma 2.5.

Let X be a compact metric space, and let $f : X \to X$ be a homeomorphism. For any set $U \subset X$, the *maximal invariant set* of f in U is defined to be

$$M(U, f) = \bigcap_{n\in\mathbb{Z}} f^n(U).$$

Thus $M(U, f)$ consists of points whose orbit never goes out of U.

A compact invariant set Λ of f is called *isolated*, or *locally maximal*, if there is a neighborhood U of Λ in X such that $M(U, f) = \Lambda$. In this case U is called an *isolating neighborhood* of Λ. Note that in this case

$$\Lambda = \bigcap_{n\in\mathbb{Z}} f^n(K)$$

for any compact set K with $\Lambda \subset \text{int}(K) \subset K \subset U$. Thus the choice of isolating neighborhood U is flexible. It can be chosen only slightly larger than Λ, either open or compact.

By Theorem 2.16, any hyperbolic fixed point is an isolated invariant set. (For instance, a hyperbolic fixed point in the horseshoe is an isolated invariant set, though it is accumulated by other periodic points.) By construction, the Smale horseshoe defined in Section 3.2 is an isolated invariant set with the square Q an isolating neighborhood.

An important example of nonisolated invariant sets is the (closure of the) orbit of a transverse homoclinic point. See Figure 4.6, where p is a hyperbolic fixed point and x is a transverse homoclinic point of p. Note that $\Lambda = \text{Orb}(x) \cup \{p\}$ is a hyperbolic set (Exercise 4.2). For any $\epsilon > 0$, there is m sufficiently large such that both $f^{-m}x$ and $f^m x$ are close to p such that $W^s_\epsilon(f^{-m}x)$ and $W^u_\epsilon(f^m x)$ intersect at a point z. Then the orbit of z is in the ϵ-neighborhood of Λ, but not in Λ. Thus Λ is not isolated.

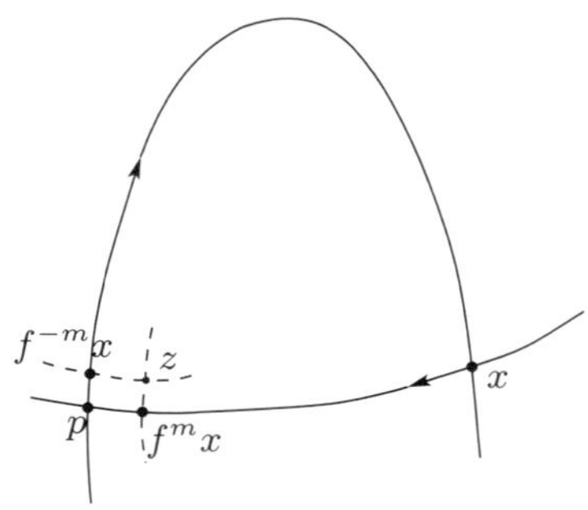

Figure 4.6. Nonisolated invariant set Λ. Under iteration, positive or negative, the point z is always close to but not in Λ.

It is interesting to note that, while every compact invariant set can be extended to an isolated invariant set (for instance to the whole manifold M), Crovisier (2001) observed that not every hyperbolic set can be extended to an isolated hyperbolic set. See also Fisher (2006).

Theorem 4.22. *Let $f : X \to X$ be a homeomorphism. Let $\Lambda \subset X$ be an isolated invariant set of f with an isolating neighborhood U. For any $a > 0$, there is a C^0 neighborhood $\mathcal{U}$ of f such that, for any $g \in \mathcal{U}$, the maximal invariant set of g in U is contained in $B(\Lambda, a)$.*

Proof. Since

$$\bigcap_{n\in\mathbb{Z}} f^n(U) = \Lambda,$$

for any $a > 0$, there is N large such that

$$\bigcap_{n=-N}^{N} f^n(U) \subset B(\Lambda, a/2).$$

Take a sufficiently small C^0 neighborhood $\mathcal{U}$ of f such that, for any $g \in \mathcal{U}$,

$$\bigcap_{n=-N}^{N} g^n(U) \subset B(\Lambda, a).$$

Then

$$\bigcap_{n=-\infty}^{\infty} g^n(U) \subset B(\Lambda, a).$$

□

The next theorem will play a crucial role in the proof of the celebrated Ω-stability theorem of Smale (Theorem 5.8).

Theorem 4.23 (Structural stability of isolated hyperbolic sets). *Let Λ be an isolated hyperbolic set of $f : M \to M$ with an isolating neighborhood U. For any $\epsilon > 0$, there is a C^1 neighborhood $\mathcal{U}$ of f such that, for any $g \in \mathcal{U}$, the maximal invariant set Γ of g in U is isolated in U and $g|_\Gamma$ is ϵ-conjugate to $f|_\Lambda$.*

This result is different from Theorem 4.19 or 4.20. The invariant set $h(\Lambda)$ of g in the definition of the ϵ-embedding stability depends on h and hence depends on both f and g. Here the set Γ is specified by the isolating neighborhood U and hence depends on g only.

Proof. By Theorem 4.21, there are a C^1 neighborhood $\mathcal{U}_0$ of f and two numbers $a_0 > 0$ and $\epsilon_0 > 0$ such that, for any $g, g' \in \mathcal{U}_0$ and any compact invariant set $\Delta \subset B(\Lambda, a_0)$ of g, there is at most one continuous map $h : \Delta \to M$ such that

$$hg = g'h, \quad d(h, id) \le \epsilon_0.$$

We may assume

$$B(\Lambda, a_0 + \epsilon_0) \subset U.$$

By Theorem 4.22, there is a C^0 neighborhood $\mathcal{U}_1$ of f such that, for any $g \in \mathcal{U}_1$, the maximal invariant set Γ of g in U satisfies

$$\Gamma \subset B(\Lambda, a_0).$$

Let $\epsilon > 0$ be given. We may assume $\epsilon \le \epsilon_0/2$. By Theorem 4.21, there is a C^1 neighborhood $\mathcal{U}$ of f with

$$\mathcal{U} \subset \mathcal{U}_0 \cap \mathcal{U}_1$$

such that, for any $g, g' \in \mathcal{U}$ and any compact invariant set $\Delta \subset B(\Lambda, a_0)$ of g, there is at least one continuous injective map $h : \Delta \to M$ such that

$$hg = g'h, \quad d(h, id) \le \epsilon.$$

In particular, since

$$\Gamma \subset B(\Lambda, a_0),$$

there is a continuous injective map $h : \Gamma \to M$ onto its image with

$$hg = g'h, \quad d(h,\ id) \le \epsilon \text{ on } \Gamma.$$

Likewise, since the maximal invariant set Γ' of g' in U satisfies

$$\Gamma' \subset B(\Lambda, a_0),$$

there is a continuous injective map $h' : \Gamma' \to M$ with

$$h'g' = gh', \quad d(h',\ id) \le \epsilon \text{ on } \Gamma'.$$

Since

$$h(\Gamma) \subset B(\Gamma, \epsilon) \subset B(\Lambda, a_0 + \epsilon) \subset B(\Lambda, a_0 + \epsilon_0) \subset U$$

and since $h(\Gamma)$ is g'-invariant and Γ' is the maximal invariant set of g' in U, it follows that

$$h(\Gamma) \subset \Gamma'.$$

Likewise,

$$h'(\Gamma') \subset \Gamma.$$

Combing h and h' together gives $h'h : \Gamma \to \Gamma$ with

$$(h'h)g = g(h'h), \quad d(h'h, id|_\Gamma) \leq \epsilon + \epsilon \leq \epsilon_0 \text{ on } \Gamma.$$

However, $id|_\Gamma$ is already an ϵ_0-conjugacy between $g|_\Gamma$ and itself. By the uniqueness stated in the beginning of the proof,

$$h'h = id|_\Gamma.$$

Likewise,

$$hh' = id|_{\Gamma'}.$$

This proves $h(\Gamma) = \Gamma'$. Hence $g|_\Gamma$ and $g'|_{\Gamma'}$ are ϵ-conjugate. Letting $g' = f$ and $\Gamma' = \Lambda$ then proves Theorem 4.23. □

4.6. The shadowing lemma

In this section we introduce the shadowing lemma of Anosov, a result that has many important applications. Let $\delta > 0$. We call a sequence $\{x_n\}_{n=-\infty}^{\infty}$ in M a *δ-pseudo-orbit* of f if, for every $n \in \mathbb{Z}$,

$$d(f(x_n),\ x_{n+1}) < \delta.$$

Thus a δ-pseudo-orbit is just a bi-infinite δ-chain (see the terminology in Section 1.1). We say a point $y \in M$ *ϵ-shadows* a pseudo-orbit $\{x_n\}_{n=-\infty}^{\infty}$ if, for every $n \in \mathbb{Z}$,

$$d(f^n(y),\ x_n) < \epsilon.$$

If $\delta \leq \delta_0$, then a δ-pseudo-orbit is automatically a δ_0-pseudo-orbit. If $\epsilon \leq \epsilon_0$, then to be ϵ-shadowed by a point is automatically to be ϵ_0-shadowed by this point.

Theorem 4.24 (The shadowing lemma). *Let $\Lambda \subset M$ be a hyperbolic set of f. Then there are $\epsilon_0 > 0$ and $\delta_0 > 0$ such that every δ_0-pseudo orbit in Λ can be ϵ_0-shadowed by at most one point. Moreover, for any $0 < \epsilon \leq \epsilon_0$, there is $0 < \delta \leq \delta_0$ such that every δ-pseudo orbit in Λ is ϵ-shadowed by at least one point.*

Note that the uniqueness can be derived directly from the expansivity of hyperbolic sets.

The shadowing lemma is usually and beautifully proved by transferring it into a fixed point problem of a Banach space; see Katok-Hasselblatt (1995) or Pilyugin (1999). Here we just explain that the shadowing lemma can be regarded as an invariant section problem as well. The idea is very intuitive. Assume a pseudo-orbit $\{x_n\}$ is ϵ-shadowed by a point $y \in M$; that is,

$$d(f^n(y),\ x_n) < \epsilon, \quad \forall\ n \in \mathbb{Z}.$$

Briefly, $f^n(y) - x_n$ is a section based on the pseudo-orbit $\{x_n\}$ with size no greater than ϵ. Define a fiber-preserving map F on vectors at $\{x_n\}$ by taking the vector $z - x_n$ to the vector $fz - x_{n+1}$. Intuitively, it applies f to the "top" z of the vector and the "shift" map to the "end" x_n of the vector. Then F is fiber-preserving over the shift map, and the section $f^n(y) - x_n$ is just an invariant section of F. The existence of a shadowing point is just the existence of such an invariant section.

There is a delicate point here that needs attention: a pseudo-orbit $\{x_n\}_{n=-\infty}^{\infty}$ is a sequence but not a set; hence it could happen that $x_n = x_m$ but $f^n(y) \neq f^m(y)$. In that case we would have defined two different vectors $f^n(y) - x_n$ and $f^m(y) - x_n$ at the same point x_n, and the section would not be well defined. Thanks to C. S. Lin for pointing this out to me. To overcome this, we pull back the pseudo-orbit $\{x_n\}_{n=-\infty}^{\infty}$ to $\mathbb{Z}$.

Precisely, given a sequence

$$\chi : \mathbb{Z} \to M$$
$$\chi(n) = x_n,$$

we define the *induced bundle* (or *pull-back bundle*) of χ to be

$$\chi^*(TM) = \{(n, v) \in \mathbb{Z} \times TM \mid v \in T_{x_n}M\},$$

where the base space is $\mathbb{Z}$ with the discrete topology and the bundle projection is

$$\pi(n,\ v) = n.$$

Thus the fiber of $\chi^*(TM)$ at n is just $(n, T_{x_n}M)$. This is like sticking a label n onto the original fiber $T_{x_n}M$ so that even if $x_n = x_m$, the new fibers $(n, T_{x_n}M)$ and $(m, T_{x_m}M)$ are still distinct, solving the above problem.

Since the base space $\mathbb{Z}$ is endowed with the discrete topology, the induced bundle $\chi^*(TM)$ is quite simple and is a discrete union of countably many Euclidean spaces. Thus, a *bundle isomorphism* $A : \chi^*(TM) \to \chi^*(TM)$ over the shift map $n \to n+1$ on $\mathbb{Z}$ is just a bi-sequence A_n of linear isomorphisms of a Euclidean space to a Euclidean space. Also, a *hyperbolic bundle isomorphism* $A : \chi^*(TM) \to \chi^*(TM)$ over the shift map is just a hyperbolic sequence.

Recall that Lemma 4.18 says that a fiber-Lipschitz perturbation of a hyperbolic bundle isomorphism has a unique invariant section near the 0-section. The corresponding result for the induced bundle $\chi^*(TM)$ is Lemma 4.25 below. Denote by Γ_χ the space of bounded sections of $\chi^*(TM)$; that is,

$$\Gamma_\chi = \{\gamma : \mathbb{Z} \to TM \mid \gamma \text{ is bounded, and } \gamma(n) \in T_{x_n}M\}.$$

Γ_χ is a Banach space with respect to the norm

$$|\gamma| = \sup_{n\in\mathbb{Z}}\{|\gamma(n)|\}.$$

Lemma 4.25. *Let $\chi : \mathbb{Z} \to M$ be a bi-sequence, and let $A : \chi^*(TM) \to \chi^*(TM)$ be a hyperbolic bundle isomorphism over the shift map $n \to n+1$ with skewness $0 < \tau < 1$ with respect to a C^0 norm $|\cdot|$ that is adapted to and of box type to the hyperbolic splitting. Let $r > 0$. If $\phi : \chi^*(TM)(r) \to \chi^*(TM)$ is fiber-preserving over $n \to n+1$ and fiber-Lipschitz with*

$$\text{Lip}_2\phi < 1 - \tau,$$

then $A + \phi$ has at most one invariant section in $\Gamma_\chi(r)$. If, in addition,

$$|\phi(0_\chi)| = \sup_{n\in\mathbb{Z}}|\phi(0_{x_n})| \le (1 - \tau - \text{Lip}_2\phi)r,$$

then $A + \phi$ has at least one (hence a unique) invariant section γ_ϕ in $\Gamma_\chi(r)$. Moreover,

$$|\gamma_\phi| \le \frac{|\phi(0_\chi)|}{1 - \tau - \text{Lip}n_2\phi}.$$

The proof of Lemma 4.25 is exactly the same as that of Lemma 4.18 and hence is omitted. Now we prove Theorem 4.24.

Proof. Take $\epsilon_0 > 0$ and $\delta_0 > 0$ small enough so that, for every δ_0-pseudo-orbit $\chi = \{x_n\}_{n=-\infty}^{\infty}$ in Λ, the *lift of f along χ*

$$F = F_\chi^f : \chi^*(TM)(\epsilon_0) \to \chi^*(TM)$$

$$F(n, v) = (n+1,\ exp_{x_{n+1}}^{-1}\ f\ exp_{x_n}(v))$$

can be defined. Clearly, F is fiber-preserving over the shift map $n \to n+1$, taking the vector $(n,\ exp_{x_n}^{-1}(z))$ to the vector $(n+1,\ exp_{x_{n+1}}^{-1}(fz))$. In particular, for $0 < \epsilon \le \epsilon_0$, if a point $y \in M$ satisfies

$$d(f^n(y),\ x_n) < \epsilon, \quad \forall\, n \in \mathbb{Z},$$

hence gives rise to a section

$$\gamma(n) = (n,\ exp_{x_n}^{-1}(f^n y))$$

of $\Gamma_\chi(\epsilon)$, then

$$F(\gamma(n)) = \gamma(n+1), \quad \forall\, n \in \mathbb{Z}.$$

In other words, γ is an invariant section of F. Therefore, the existence of a point y that ϵ-shadows the pseudo-orbit χ is equivalent to the existence of an invariant section γ of F in $\Gamma_\chi(\epsilon)$. The rest of the proof is just a direct consequence of Lemma 4.25 and hence is omitted. $\square$

For a compact invariant set Λ of f, define

$$W^s(\Lambda) = \{x \in M \mid d(f^n x,\ \Lambda) \to 0,\ n \to -\infty\},$$
$$W^u(\Lambda) = \{x \in M \mid d(f^{-n} x,\ \Lambda) \to 0.\ n \to +\infty\}.$$

By Theorem 1.2, $x \in W^s(\Lambda)$ if and only if $\omega(x) \subset \Lambda$.

An important application of the shadowing lemma is the next theorem, which says that if a point is asymptotic to an isolated hyperbolic set, then it is asymptotic to a point of the set. It is usually referred to as the *In Phase Theorem*. It was first proved by Hirsch-Palis-Pugh-Shub (1970).

Theorem 4.26. *Let $\Lambda \subset M$ be an isolated hyperbolic set of f. Then*

$$W^s(\Lambda) = \bigcup_{x\in\Lambda} W^s(x),$$
$$W^u(\Lambda) = \bigcup_{x\in\Lambda} W^u(x).$$

Proof. It suffices to prove

$$W^s(\Lambda) \subset \bigcup_{x\in\Lambda} W^s(x).$$

Let $y \in W^s(\Lambda)$. Let $r > 0$ be the number in Theorem 4.13 such that, for every $x \in \Lambda$,

$$W^s_r(x) = \{y \in M \mid d(f^n y, f^n x) \le r,\ \forall\ n \ge 0\}.$$

Shrinking r if necessary, we may assume that $B(\Lambda, r)$ is contained in an isolating neighborhood of Λ. By Theorem 4.24, there is $\delta > 0$ such that every δ-pseudo-orbit in Λ is $r/2$-shadowed. Take $0 < \eta < \delta$ such that if

$$d(f^n y,\ x_n) < \eta, \quad \forall\ n \in \mathbb{Z},$$

then $\{x_n\}$ is a δ-pseudo-orbit.

Since $y \in W^s(\Lambda)$, we may take $m > 0$ big enough so that

$$d(f^n y,\ \Lambda) < \eta$$

for all $n \ge m$; hence there are $x_n \in \Lambda$ such that

$$d(f^n y,\ x_n) < \eta$$

for all $n \ge m$. Then $\{x_n\}_{n=m}^\infty$ is (a piece of) a δ-pseudo-orbit. For every $n < m$, simply let

$$x_n = f^{n-m}(x_m).$$

Then $\{x_n\}_{n=-\infty}^{\infty}$ is a δ-pseudo-orbit in Λ and hence is $r/2$-shadowed by some point x. This implies

$$\mathrm{Orb}(x) \subset B(\Lambda, r/2).$$

Since Λ is isolated, it follows that $x \in \Lambda$. Clearly, for every $n \geq m$,

$$d(f^n y,\ f^n x) \leq d(f^n y,\ x_n) + d(x_n,\ f^n x) \leq \eta + r/2 \leq r.$$

By Theorem 4.13, $f^m y \in W_r^s(f^m x)$. Then $y \in W^s(x)$. This proves Theorem 4.26. □

A "perturbation" of a pseudo-orbit is a pseudo-orbit. Precisely, for any $\delta > 0$, there is $\eta > 0$ such that, for any η-pseudo-orbit $\{x_n\}$, if

$$d(y_n,\ x_n) < \eta,$$

then $\{y_n\}$ is a δ-pseudo-orbit (Exercise 1.9). Thus the pseudo-orbit $\{x_n\}$ in Theorem 4.24 does not really have to be in Λ. The following theorem gives this slight but important improvement.

Theorem 4.27 (Improving Theorem 4.24). *Let $\Lambda \subset M$ be a hyperbolic set of f. For any $\epsilon > 0$, there is $\eta > 0$ such that every η-pseudo-orbit in the η-neighborhood of Λ is ϵ-shadowed by a point.*

Proof. For any $\epsilon > 0$, by Theorem 4.24, there is $\delta > 0$ such that every δ-pseudo-orbit in Λ is $\epsilon/2$-shadowed by a point. Take $0 < \eta \leq \epsilon/2$ such that for any η-pseudo-orbit $\{x_n\}$, if $d(y_n,\ x_n) < \eta$, then $\{y_n\}$ is a δ-pseudo-orbit. Now let $\{x_n\}$ be an η-pseudo-orbit in the η-neighborhood of Λ. For each $n \in \mathbb{Z}$, take $y_n \in \Lambda$ such that $d(y_n,\ x_n) < \eta$. Then $\{y_n\}$ is a δ-pseudo-orbit in Λ and hence is $\epsilon/2$ shadowed by a point z. Then $\{x_n\}$ is ϵ-shadowed by z. □

A typical application of the shadowing lemma is to locate periodic orbits. This is referred to as the *Anosov closing lemma*, which says that, near a "periodic pseudo-orbit" in a hyperbolic set, there must be a periodic orbit. Here a pseudo-orbit $\{x_n\}_{n=-\infty}^{\infty}$ is called *periodic* if there is $m > 0$ such that $x_n = x_{n+m}$ for every $n \in \mathbb{Z}$.

Theorem 4.28 (Anosov closing lemma). *Let $\Lambda \subset M$ be a hyperbolic set of f. For every $\epsilon > 0$, there is $\delta > 0$ such that every periodic δ-pseudo-orbit in Λ is ϵ-shadowed by a periodic point.*

Proof. By Theorem 4.24, there are $\epsilon_0 > 0$ and $\delta_0 > 0$ such that every δ_0-pseudo-orbit in Λ can be ϵ_0-shadowed by at most one point.

Let $0 < \epsilon \leq \epsilon_0$ be given. By Theorem 4.24, there is $0 < \delta \leq \delta_0$ such that every δ-pseudo-orbit in Λ is ϵ-shadowed by at least one point.

Let $\{x_n\}_{n=-\infty}^{\infty}$ be a periodic δ-pseudo-orbit in Λ such that, for some $m > 0$, $x_n = x_{n+m}$ for all $n \in \mathbb{Z}$. Then $\{x_n\}$ is ϵ-shadowed by a point p; that is,

$$d(f^n p,\ x_n) < \epsilon, \quad \forall\, n \in \mathbb{Z}.$$

That is,

$$d(f^{n+m} p,\ x_{n+m}) < \epsilon, \quad \forall\, n \in \mathbb{Z}.$$

But $x_n = x_{n+m}$; hence

$$d(f^{n+m} p,\ x_n) < \epsilon, \quad \forall\, n \in \mathbb{Z}.$$

In other words, $f^m p$ also ϵ-shadows $\{x_n\}$. By the uniqueness stated in the beginning of the proof, $f^m p = p$, meaning p is periodic. This proves Theorem 4.28. □

Next is an improvement of the previous theorem, exactly like Theorem 4.27. We omit the proof.

Theorem 4.29 (Improving Theorem 4.28). *Let $\Lambda \subset M$ be a hyperbolic set of f. For every $\epsilon > 0$, there is $\eta > 0$ such that every periodic η-pseudo-orbit in the η-neighborhood of Λ is ϵ-shadowed by a periodic point.*

Remark. Since a periodic δ-pseudo-orbit is just a δ-chain repeated infinitely many times in both positive and negative directions, we may state the Anosov closing lemma by saying that every periodic δ-chain in Λ is ϵ-shadowed by a periodic point. Likewise for Theorem 4.29.

Here are several applications of the Anosov closing lemma.

Theorem 4.30. *Every transverse homoclinic point is the limit of periodic points.*

Proof. We take the case that p is a hyperbolic fixed point. Let x be a transverse homoclinic point of p. Then

$$\Lambda = \mathrm{Orb}(x) \cup \{p\}$$

is a hyperbolic set (Exercise 4.2). For any $\epsilon > 0$, let $\delta > 0$ be the constant of Λ guaranteed by the Anosov closing lemma. Take $m > 0$ big enough so that $d(f^{-m}x,\ f^m x) < \delta$. Then

$$f^{-m}x, \ldots, x, \ldots, f^{m-1}x, f^{-m}x$$

is a periodic δ-chain in Λ, going through x. By the Anosov closing lemma, it is ϵ-shadowed by a periodic point. □

Recall that $\mathrm{L}(f)$ and $\mathrm{CR}(f)$ denote the limit set and chain recurrent set of f, respectively. See the definitions in Chapter 1.

Theorem 4.31. *If $\mathrm{L}(f)$ is hyperbolic, then $\mathrm{L}(f) = \overline{\mathrm{P}(f)}$.*

Proof. It suffices to prove $\omega(x) \subset \overline{\mathrm{P}(f)}$ for every $x \in X$. Let $y \in \omega(x)$. Let $\epsilon > 0$ be given. For $\epsilon/2$, take $\eta > 0$ guaranteed by Theorem 4.29 (treating $\mathrm{L}(f)$ as Λ). We may assume $\eta < \epsilon$. There are $n_1 < n_2$ large such that $f^{n_1}(x)$ and $f^{n_2}(x)$ are both in $B(y, \eta/2)$ and the orbit-arc from $f^{n_1}(x)$ to $f^{n_2}(x)$ is entirely within the η-neighborhood of $\omega(x)$. Then

$$f^{n_1}(x), f^{n_1+1}(x), \ldots, f^{n_2-1}(x), f^{n_1}(x)$$

is a periodic η-chain in the η-neighborhood of $\omega(x)$ going through $B(y, \eta/2)$. By Theorem 4.29, it is $\epsilon/2$-shadowed by a periodic point p of f. Hence $d(y,p) < \epsilon/2 + \eta/2 < \epsilon$, proving Theorem 4.31. □

Theorem 4.32. *If* $\mathrm{CR}(f)$ *is hyperbolic, then* $\mathrm{CR}(f) = \overline{\mathrm{P}(f)}$.

Proof. Let $x \in \mathrm{CR}(f)$. For any $\epsilon > 0$, let $\eta > 0$ be the constant guaranteed by Theorem 4.29 (treating $\mathrm{CR}(f)$ as Λ). By Theorem 1.7, there is $\delta > 0$ such that any periodic δ-chain P_δ through x is contained in the η-neighborhood of $\mathrm{CR}(f)$. We may assume $\delta < \eta$. By Theorem 4.29, P_δ is ϵ-shadowed by a periodic point p of f. Hence $d(x,p) < \epsilon$, proving Theorem 4.32.

□

Exercises

As usual, M denotes a compact manifold without boundary.

Exercise 4.1. Verify that the horseshoe set Λ defined in Section 3.2 is a hyperbolic set.

Exercise 4.2. Prove that if $x \in M$ is a transverse homoclinic point of a hyperbolic fixed point $p \in M$, then $\mathrm{Orb}(x)$ is a hyperbolic orbit.

Exercise 4.3. Let $\Lambda \subset M$ be a compact invariant set of f, and let $E \subset T_\Lambda M$ be a Tf-invariant C^0 subbundle. Prove that the following three conditions are equivalent:

(a) There are $C \geq 1$ and $0 < \lambda < 1$ such that

$$|Tf^n(v)| \leq C\lambda^n |v|, \ \forall\, v \in E, n \geq 0.$$

(b) There are $0 < \mu < 1$ and $N \geq 0$ such that

$$|Tf^n(v)| \leq \mu |v|, \ \forall\, v \in E, n \geq N.$$

(c) For any $0 \neq v \in E$, there is $n = n(v) \geq 0$ such that

$$|Tf^n(v)| < |v|.$$

Exercise 4.4. Let $p \in M$ be a contracting periodic point of f. Prove that if $p \in \alpha(x)$, then $x \in \mathrm{Orb}(p)$.

Exercise 4.5. Let $\Lambda \subset M$ be a set. For any $x \in \Lambda$, let $E(x) \subset T_xM$ be an m-dimensional linear subspace. Prove $E = \bigcup_{x\in\Lambda} E(x)$ is an m-dimensional C^0 subbundle of $T_\Lambda M$ if and only if $E(x)$ varies continuously in $x \in \Lambda$.

Exercise 4.6. Let $\Lambda \subset M$ be a compact invariant set of f, and let $E \subset T_\Lambda M$ be a C^0 subbundle. Assume $F : E \to E$ is continuous and fiber-preserving over $f : \Lambda \to \Lambda$ and assume F is contracting on fibers in the sense that there is $0 < \lambda < 1$ such that for any $u, v \in E$ with $\pi u = \pi v$,

$$|F(u) - F(v)| \le \lambda |u - v|.$$

Prove there is a unique continuous section $\gamma \in \Gamma^0(E)$ that is invariant under F; that is,

$$F\gamma(x) = \gamma(fx), \ \forall\, x \in \Lambda.$$

Exercise 4.7. Let $\Lambda \subset M$ be a compact invariant set of f. Assume that, with respect to a C^0 direct sum

$$T_\Lambda M = E_1 \oplus E_2,$$

$Tf = A$ is represented as

$$\begin{pmatrix} A_{11} & 0 \\ A_{21} & A_{22} \end{pmatrix}$$

such that

$$\max\{|A_{11}^{-1}|, |A_{22}|\} < 1.$$

Prove that Λ is a hyperbolic set of f.

Exercise 4.8. Prove Lemma 4.7.

Exercise 4.9. Prove there is no expansive homeomorphism on the unit interval. How about on the circle?

For the statement of some exercises below we insert a definition here. Let $\Lambda \subset M$ be a compact invariant set of f. A continuous invariant splitting $T_xM = G^s(x) \oplus G^u(x)$, $x \in \Lambda$, is called *dominated* if there are $C \ge 1$ and $\lambda \in (0, 1)$ such that

$$|Tf^n|_{G^s(x)}| \cdot |Tf^{-n}|_{G^u(f^n(x))}| \le C\lambda^n$$

for any $x \in \Lambda$ and $n \in \mathbb{N}$ or, in other words, if

$$\frac{|Tf(v_s)|}{|Tf(v_u)|} \le C\lambda^n$$

for any $x \in \Lambda$, any $v_s \in G^s(x)$ with $|v_s| = 1$ and any $v_u \in G^u(x)$ with $|v_u| = 1$, and any $n \in \mathbb{N}$. Obviously, a hyperbolic splitting is dominated.

Exercise 4.10. Let $T_\Lambda M = G_1^s \oplus G_1^u$ and $T_\Lambda M = G_2^s \oplus G_2^u$ be two dominated splittings on Λ. Show that for any $x \in \Lambda$, either $G_1^s(x) \subseteq G_2^s(x)$ or $G_1^u(x) \subseteq G_2^u(x)$. In particular, if $\dim G_1^s(x) = \dim G_2^s(x)$, then $G_1^s(x) = G_2^s(x)$ and $G_1^u(x) = G_2^u(x)$ (that is, for fixed index, dominated splitting is unique).

Exercise 4.11. Let $T_\Lambda M = G_1^s \oplus G_1^c \oplus G_1^u$ and $T_\Lambda M = G_2^s \oplus G_2^c \oplus G_2^u$ be three-way dominated splittings on Λ. Assume $G_1^c = G_2^c$. Prove that $G_1^s = G_2^s$ and $G_1^u = G_2^u$. (A three-way splitting $G^s \oplus G^c \oplus G^u$ is called *dominated* if G^s is dominated by G^c and G^c is dominated by G^u.)

Exercise 4.12. Let $\Lambda \subset M$ be a compact invariant set of f with a dominated splitting. Prove there are a C^1 neighborhood $\mathcal{U}$ of f and a number $a > 0$ such that for any $g \in \mathcal{U}$, any compact invariant set $\Delta \subset B(\Lambda, a)$ of g has a dominated splitting with respect to g.

Exercise 4.13. Assume $\dim M = 2$. Let $TM = G^s \oplus G^u$ be a dominated splitting of f. Prove there is a C^1 neighborhood $\mathcal{U}$ of f such that any $g \in \mathcal{U}$ has no periodic orbit with nonreal eigenvalue.

Exercise 4.14. Assume $\dim M = 3$. Let Λ be a transitive set of f that contains two hyperbolic periodic points p and q, such that p has a nonreal eigenvalue of norm less than 1 and $\dim W^s(p) = 2$ and such that q has a nonreal eigenvalue of norm greater than 1 and $\dim W^s(q) = 1$. Prove that f has no dominated splitting on $T_\Lambda M$.

Exercise 4.15. Let $TM = G^s \oplus G^u$ be a dominated splitting of f. Let $p \in M$ and $q \in M$ be two hyperbolic periodic saddles of f with $\dim W^s(p) = \dim W^s(q) = \dim G^s$. Prove $W^u(p)$ intersects $W^s(q)$ transversely.

Exercise 4.16. Let $T_\Lambda M = G^s \oplus G^u$ be a dominated splitting on Λ. Prove if $\dim G^u = 1$, there exists a norm $|\cdot|$ on $T_\Lambda M$ (said to be *adapted* to Λ), such that

$$|Tf|_{G^s(x)}| \cdot |Tf^{-1}|_{G^u(f(x))}| < 1, \ \forall x \in \Lambda.$$

Exercise 4.17. Let $\Lambda \subset M$ be a hyperbolic set. Prove there is $\epsilon > 0$ such that for any $x, y \in \Lambda$, the intersection $W^s_\epsilon(x) \cap W^u_\epsilon(y)$ consist of at most one point and that there is $\delta > 0$ such that whenever $d(x, y) < \delta$, $x, y \in \Lambda$, then $W^s_\epsilon(x) \cap W^u_\epsilon(y) \neq \emptyset$.

Exercise 4.18. Let $\Lambda \subset M$ be an isolated hyperbolic set of f. Prove there is $\epsilon > 0$ such that $W^s_\epsilon(x) \cap W^u_\epsilon(y) \in \Lambda$ for any $x, y \in \Lambda$.

Exercise 4.19. Let $0 = (\ldots, 0, 0, 0, \ldots) \in \Sigma_2$. Let $a \in \Sigma_2$ be the point such that $a_0 = 1$ but $a_n = 0$ for all $n \neq 0$. Let $\Lambda = \{0\} \cup \mathrm{Orb}(a)$. Prove Λ is compact invariant but not isolated with respect to the shift map σ.

Exercise 4.20. Let $\Lambda \subset M$ be a hyperbolic set of f with splitting $T_\Lambda M = E^s \oplus E^u$, and let $E \subset T_\Lambda M$ be a 1-dimensional Tf-invariant C^0 subbundle. If Λ is transitive, prove that either $E \subset E^s$ or $E \subset E^u$.

Exercise 4.21. Let $f : M \to M$ be an Anosov diffeomorphism, and let $\Gamma \subset M$ be a smooth simple closed curve. Prove that Γ cannot be f-invariant.

Exercise 4.22. Let $T_xM = E(x) \oplus F(x)$, $x \in M$, be a continuous invariant splitting of f. Prove that if this splitting restricted to the nonwandering set $\Omega(f)$ is hyperbolic, then the whole splitting is hyperbolic (f is Anosov).

Exercise 4.23. Let $\Omega(f)$ be hyperbolic. Prove $\Omega(f|_{\Omega(f)}) = \overline{\mathrm{P}(f)}$.

Exercise 4.24. Let $B \subset M$ be a transitive set of f. Assume $\omega(x) \subset B$ and $\alpha(x) \subset B$. Prove for any $\delta > 0$ that there is a periodic δ-chain in $\mathrm{Orb}(x) \cup B$ from x to x.

A hyperbolic set Λ of f is said to have the *local product structure* if for $r > 0$ small enough, there is $\delta > 0$ such that, for any pair of points $x, y \in \Lambda$ with $d(x, y) < \delta$, the intersection $W_r^s(x) \cap W_r^u(y)$ is a single point which is in Λ.

Exercise 4.25. Let $\Lambda \subset M$ be a hyperbolic set of f. Prove f has local product structure if and only if Λ is isolated.

Exercise 4.26. Let $f : M \to M$ be an Anosov diffeomorphism. Prove that if the unstable manifold $W^u(x)$ of every point $x \in M$ is dense in M, then f is mixing.

Exercise 4.27. Let $f : M \to M$ be an Anosov diffeomorphism with $\Omega(f) = M$. We also assume M is connected. Prove the following:

(1) For every periodic point p of f, $W^s(p)$ is dense in M.

(2) For every point $x \in M$, $W^s(x)$ is dense in M.

(Hence f is mixing by the previous exercise.)

Exercise 4.28. Let $p \in M$ be a hyperbolic fixed point of f. Assume there is a sequence of periodic points p_k of f with $p_k \neq p$ but $p_k \to p$. Prove there are sequences $k_i \to \infty$ and $m_{k_i} \to \infty$ such that $f^{m_{k_i}}(p_{k_i})$ converge to a point on $W^u(p) - \{p\}$.

Exercise 4.29. Let $f : M \to M$ be a C^r structurally stable diffeomorphism, $r \geq 1$. Prove every periodic point of f is hyperbolic.

Chapter 5

Axiom A, no-cycle condition, and Ω-stability

In this chapter we investigate the structural stability of the nonwandering set $\Omega(f)$, proving the celebrated Ω-stability theorem of Smale (1970). Then we re-examine the issue in terms of the limit set $\mathrm{L}(f)$ and the chain recurrent set $\mathrm{CR}(f)$, based on ideas of Newhouse (1972, 1980) and Franke-Selgrade (1977), respectively.

5.1. Spectral decomposition and Axiom A

We start with the λ-lemma of Palis, a tool that highlights a crucial point of chaotic dynamics. For a hyperbolic fixed point $p \in M$ of f, let us abbreviate $u = \dim W^u(p)$ and call a u-dimensional C^1 embedded disc in M a *u-disc.* Likewise for *s-disc.*

Theorem 5.1 (The λ-lemma). *Let $p \in M$ be a hyperbolic fixed point of $f : M \to M$. For any u-disc B in $W^u(p)$, any point $x \in W^s(p)$, any u-disc D transverse to $W^s(p)$ at x, and any $\epsilon > 0$, there is $N > 0$ such that if $n > N$, $f^n(D)$ contains a u-disc that is C^1 ϵ-close to B*

To visualize the statement see Figure 5.1. Briefly, the λ-lemma states that no matter how big B is, how small D is, and how weakly transverse (small angle) D to $W^s(p)$ is, the conclusion is always true. The λ-lemma holds similarly for s-discs. The proof can be found in many textbooks, for instance Palis and de Melo (1982).

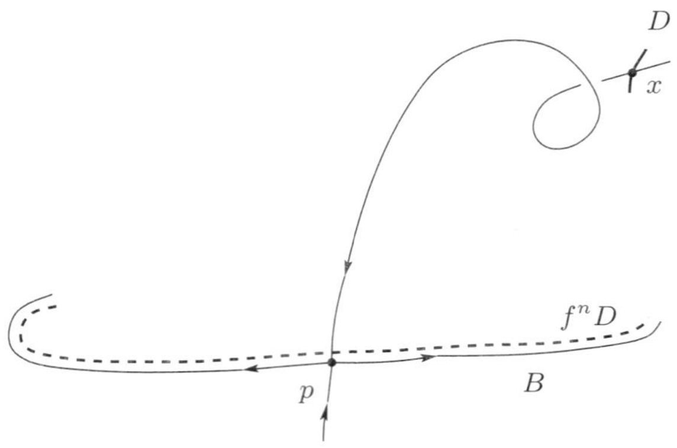

Figure 5.1. The statement of the λ-lemma.

The λ-lemma is very useful in applications. We give some typical examples to illustrate this.

Example 1. Consider a 2D diffeomorphism f that has a homoclinic loop associated with a hyperbolic fixed point p. See Figure 5.2. Using the λ-lemma we conclude that *every point x on the loop is nonwandering.* This is because every small arc D transverse to the loop at x will eventually pile on $W^u(p)$, hence to cross D itself.

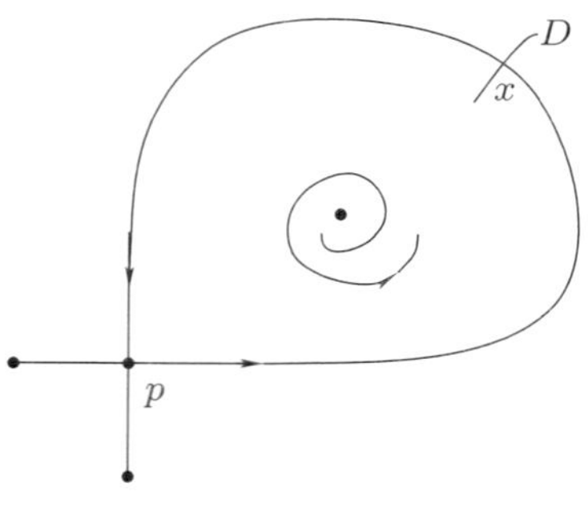

Figure 5.2. A homoclinic loop.

Example 2. Let p and q be two hyperbolic periodic points of f such that $W^s(\text{Orb}(p))$ and $W^u(\text{Orb}(q))$ have a transverse intersection and $W^s(\text{Orb}(q))$ and $W^u(\text{Orb}(p))$ have a transverse intersection. See Figure 5.3. Then *every point of the transverse intersection is nonwandering.* For instance we verify that $x \in W^s(\text{Orb}(p)) \pitchfork W^u(\text{Orb}(q))$ is nonwandering. Switching to an iterate of f if necessary, we may assume p and q are fixed points of f. Take a small disc D in $W^u(q)$ of center x and of dimension $\dim(W^u(q))$. By the λ-lemma, iterates of D will pile on $W^u(p)$ and hence cross $W^s(q)$ to give a small disc D_1 of center $z \in W^s(q)$ and of dimension $\dim(W^u(q))$. Then by the λ-lemma, iterates of D_1 will pile on $W^u(q)$ to get back near x. Thus $x \in \Omega(f)$.

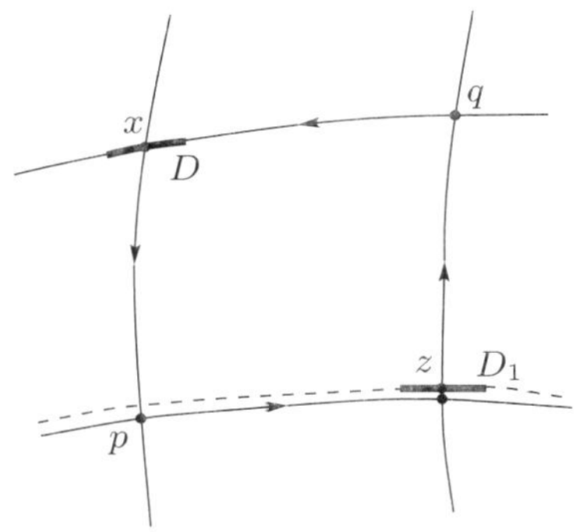

Figure 5.3. Homoclinically related periodic points.

Using the Anosov closing lemma one can obtain a sharper conclusion: $x \in \overline{P(f)}$. See the argument in the proof of the spectral decomposition theorem below.

Hyperbolicity is the key to understanding structural stability. First we put the hyperbolicity assumption on $\overline{P(f)}$, the closure of periodic points. A striking feature derived then will be a "finiteness" described in the following theorem due to Smale (1967).

Theorem 5.2 (The spectral decomposition theorem). *Let $f : M \to M$ be a diffeomorphism such that $\overline{P(f)}$ is hyperbolic. Then $\overline{P(f)}$ decomposes in a unique way into finitely many disjoint transitive sets*

$$\overline{P(f)} = B_1 \cup \cdots \cup B_k.$$

Proof. First we prove the uniqueness of the decomposition. Assume there is another decomposition

$$\overline{P(f)} = C_1 \cup \cdots \cup C_l$$

of the same property. For each $i = 1, \ldots, k$, take $x_i \in B_i$ such that $\omega(x_i) = B_i$. There is a unique j with $x_i \in C_j$. Hence $B_i \subset C_j$. Since $\bigcup B_i = \bigcup C_j$, it follows that $k \geq l$. Similarly, $l \geq k$. Thus $k = l$. Again, since $\bigcup B_i = \bigcup C_j$, the above inclusion $B_i \subset C_j$ is actually an equality. This proves the uniqueness.

By Theorem 4.1, f has at most finitely many contracting and expanding periodic orbits. Since they are away from the other points of $\overline{P(f)}$ with a positive distance, without loss of generality we assume that all periodic points of f are of the saddle type.

Define a binary relation $\sim$ on the set $\mathrm{P}(f)$ of periodic points by $p \sim q$ if and only if $W^s(p)$ and $W^u(q)$ intersect transversely and $W^u(p)$ and $W^s(q)$ intersect transversely. Note that $p \sim q$ implies that p and q have the same index. The relation is reflective and symmetric. We prove it is transitive; that is, if $p \sim q$ and $q \sim r$, then $p \sim r$. Let $g = f^k$, where k is the product of the periods of p, q, and r. Then p, q, and r are fixed points of g. Note that $W^s(p, f) = W^s(p, g)$. Likewise for q and r and the unstable manifolds. Then $p \sim q$ and $q \sim r$ with respect to g. Applying the λ-lemma to g, it follows immediately that $p \sim r$ with respect to g. Then $p \sim r$ with respect to f. This proves that $\sim$ is an equivalence relation.

Thus $\mathrm{P}(f)$ decomposes into equivalence classes of $\sim$. By Theorem 4.17, there is $\delta > 0$ such that if $p, q \in \mathrm{P}(f)$ are within δ, then $p \sim q$. Hence any two equivalence classes are of distance at least δ. Hence there are only finitely many equivalence classes

$$P_1, P_2, \ldots, P_N$$

of $\mathrm{P}(f)$, and

$$\overline{P}_i \cap \overline{P}_j = \emptyset, \quad i \neq j.$$

Note that P_i may not be invariant. Nevertheless since f preserves equivalence classes, for any i, there is a unique j such that $f(P_i) = P_j$ and hence $f(\overline{P}_i) = \overline{P}_j$. Obviously, the map $i \to j$ is a bijection on $\{1, 2, \ldots, N\}$ and hence is a product of cyclic permutations. Then $\overline{\mathrm{P}(f)}$ decomposes into a disjoint union of compact invariant sets

$$\overline{\mathrm{P}(f)} = B_1 \cup \cdots \cup B_k,$$

each B_i being a cyclic union of some $\overline{P}_s$.

It remains to prove that each B_i is transitive. For explicitness we prove B_1 is transitive. We may assume

$$B_1 = \overline{P}_1 \cup \cdots \cup \overline{P}_r$$

such that $f(\overline{P}_1) = \overline{P}_2, \ldots, f(\overline{P}_r) = \overline{P}_1$. Take any two open sets U and V of B_1 (which are generally not open sets of M). By Theorem 1.6, it suffices to prove there is $n \geq 1$ such that $f^n(U) \cap V \neq \emptyset$. We may assume $U \subset \overline{P}_s$, $V \subset \overline{P}_t$, $1 \leq s \leq t \leq r$. Denote $W = f^{t-s}(U)$. Then $W \subset \overline{P}_t$. It suffices to prove there is $n \geq 1$ such that $f^n(W) \cap V \neq \emptyset$.

Take $p \in W \cap P_t$ and $q \in V \cap P_t$. Then $p \sim q$. That is, there are transverse intersections $z \in W^u(p) \cap W^s(q)$ and $z' \in W^u(q) \cap W^s(p)$. To prove there is $n \geq 1$ such that $f^n(W) \cap V \neq \emptyset$, replacing f by f^k if necessary, where k is the product of the periods of p and q, we may simply assume p and q are fixed points of f. Then

$$\Lambda = \{p\} \cup \{q\} \cup \mathrm{Orb}(z) \cup \mathrm{Orb}(z')$$

is a hyperbolic set. For large $m \geq 1$, $f^{-m}z, \ldots, z, \ldots, f^{m-1}z, f^{-m}z', \ldots, z', \ldots, f^{m-1}z', f^{-m}z$ is a periodic chain in Λ. By the Anosov closing lemma, $z \in \overline{\mathrm{P}(f)}$. Hence z belongs to some B_i. Since the B_i's are mutually disjoint compact invariant sets and $d(f^n z, q) \to 0$, it follows that $z \in B_1$. Then there are $m \geq 1$ and $l \geq 1$ sufficiently large such that $f^{-m}(z) \in W$ and $f^l(z) \in V$. Let $n = m + l$. Then $f^n(W) \cap V \neq \emptyset$. This proves Theorem 5.2. $\square$

For a diffeomorphism f with $\overline{\mathrm{P}(f)}$ hyperbolic, we call B_i in the decomposition a *basic set* of $\overline{\mathrm{P}(f)}$, or a *basic set* of f.

Theorem 5.3. *If $\overline{\mathrm{P}(f)}$ is hyperbolic, then $\overline{\mathrm{P}(f)}$ is isolcted.*

Proof. Let

$$\overline{\mathrm{P}(f)} = B_1 \cup \cdots \cup B_k$$

be the spectral decomposition of f. It suffices to prove that each basic set, say B_1, is isolated. Take a compact neighborhood U_1 of B_1, disjoint from the other B_i, such that the maximal invariant set Δ_1 of f in U_1 is hyperbolic. It suffices to prove $\Delta_1 \subset B_1$. Let $x \in \Delta_1$. We prove $x \in B_1$.

Since $\omega(x) \subset \Delta_1$ is hyperbolic, every point $y \in \omega(x)$ is accumulated by periodic points whose orbits are contained in U_1 (see the proof of Theorem 4.31). Thus $\omega(x) \subset B_1$. Likewise $\alpha(x) \subset B_1$. Since B_1 is transitive, for any $\delta > 0$, there is a periodic δ-chain in $\mathrm{Orb}(x) \cup B_1 \subset \Delta_1$ from x to x. (This δ-chain starts with x, gets near $\omega(x)$, jumps into B_1, and then, with the help of a dense orbit of B_1, gets near and jumps onto a point on the negative orbit of x, and then gets back to x). By the Anosov closing lemma, there are shadowing periodic points $p_n \to x$ with $\mathrm{Orb}(p_n) \subset U_1$. Thus $x \in B_1$, proving Theorem 5.3. $\square$

Generally the set $\overline{\mathrm{P}(f)}$ does not capture the limit behavior of all orbits, and in that case we need to consider the limit set $\mathrm{L}(f)$, as in the next theorem. As before, if a result is purely topological, we will state the result in a compact metric space X.

Theorem 5.4. *Let $f : X \to X$ be a homeomorphism, and let $\Lambda_1, \ldots, \Lambda_k$ be finitely many disjoint compact invariant sets of f with $\Lambda_1 \cup \cdots \cup \Lambda_k \supset \mathrm{L}(f)$. Then*

$$X = \bigcup_{i=1}^{k} W^s(\Lambda_i) = \bigcup_{i=1}^{k} W^u(\Lambda_i).$$

Proof. Take a compact neighborhood U_i of Λ_i in X such that, for any $i \neq j$,

$$U_i \cap U_j = \emptyset, \quad (fU_i) \cap U_j = \emptyset.$$

The second equality means that a point in U_i cannot jump positively into a different U_j in one step. Switching i and j gives the negative case.

Let $x \in X$. Since $\omega(x) \subset \mathrm{L}(f)$, there is $N \geq 1$ such that for all $n \geq N$, $f^n x \in \bigcup_{i=1}^k U_i$. To be precise we assume $f^N x \in U_1$. Since $f^N x$ cannot jump into any different U_j in one step, we have $f^{N+1} x \in U_1$. Inductively, $f^n x \in U_1$ for all $n \geq N$. Then $\omega(x) \subset U_1$. Hence $\omega(x) \subset \Lambda_1$, meaning $x \in W^s(\Lambda_1)$. Likewise for W^u. This proves Theorem 5.4. □

Let us strengthen the result one step further.

Theorem 5.5. *Let $f : M \to M$ be a diffeomorphism. If $\mathrm{L}(f)$ is hyperbolic, then*

$$M = \bigcup_{x \in \mathrm{L}(f)} W^s(x) = \bigcup_{x \in \mathrm{L}(f)} W^u(x).$$

Proof. Since $\mathrm{L}(f)$ is hyperbolic, by Theorem 4.31, $\mathrm{L}(f) = \overline{\mathrm{P}(f)}$. By Theorem 5.3, $\overline{\mathrm{P}(f)}$, hence $\mathrm{L}(f)$, is isolated. Since

$$M = W^s(\mathrm{L}(f)) = W^u(\mathrm{L}(f)),$$

by Theorem 4.26,

$$M = \bigcup_{x \in \mathrm{L}(f)} W^s(x) = \bigcup_{x \in \mathrm{L}(f)} W^u(x).$$

□

To study the structural stability of the nonwandering set $\Omega(f)$, Smale (1967) proposed the following condition on $\Omega(f)$: A diffeomorphism $f : M \to M$ is said to satisfy *Axiom A* if $\Omega(f)$ is hyperbolic and if $\Omega(f) = \overline{\mathrm{P}(f)}$.

The (algebraic) Anosov automorphisms on the 2-torus defined in Section 3.3 satisfy Axiom A. Generally, every Anosov diffeomorphism satisfies Axiom A (Exercise 5.23). The horseshoe map $f : S^2 \to S^2$ in Section 3.2 is naturally defined so that $\Omega(f) = \Lambda \cup \{N\} \cup \{S\}$, where N is a hyperbolic sink of f in the upper hemisphere and S is a hyperbolic source of f in the lower hemisphere. Hence f satisfies Axiom A.

On the other hand, the system given in Figure 5.2 does not satisfy Axiom A.

All the usual examples seem to hint that if $\Omega(f)$ is hyperbolic, then $\Omega(f) = \overline{\mathrm{P}(f)}$. However, Dankner (1978) found a striking example in dimension 3 such that $\Omega(f)$ is hyperbolic but $\Omega(f) \neq \overline{\mathrm{P}(f)}$. Recall that if $\mathrm{L}(f)$ is hyperbolic, then $\mathrm{L}(f) = \overline{\mathrm{P}(f)}$ (Theorems 4.31), and if $\mathrm{CR}(f)$ is hyperbolic, then $\mathrm{CR}(f) = \overline{\mathrm{P}(f)}$ (Theorem 4.32). This sounds strange. But the point is, for $\mathrm{L}(f)$ the periodic δ-chain can be chosen entirely in an arbitrarily small neighborhood of $\mathrm{L}(f)$; hence the shadowing lemma applies and yields periodic orbits. Likewise for $\mathrm{CR}(f)$. In contrast, for $\Omega(f)$, it may happen that all periodic δ-chains go out of a neighborhood of $\Omega(f)$.

The spectral decomposition theorem was originally stated under the assumption of Axiom A. Newhouse (1980) first noticed that this is not necessary and isolated the actual assumption used in the argument.

5.2. Cycle and Ω-explosion

A diffeomorphism $f : M \to M$ is called C^r *Ω-stable* if there is a C^r neighborhood $\mathcal{U}$ of f such that for any $g \in \mathcal{U}$, $g|_{\Omega(g)}$ is conjugate to $f|_{\Omega(f)}$. Specifically, f is called C^r *ϵ-Ω-stable* if for any $\epsilon > 0$, there is a C^r neighborhood $\mathcal{U}$ of f such that for any $g \in \mathcal{U}$, $g|_{\Omega(g)}$ is conjugate to $f|_{\Omega(f)}$ and the conjugacy is ϵ-close to id.

By Theorem 5.3, if f satisfies Axiom A, then $\Omega(f)$ is isolated. Hence $\Omega(f)$ has the structural stability of an isolated hyperbolic set guaranteed by Theorem 4.23. But this does not mean f is Ω-stable. In fact Axiom A alone is not sufficient for Ω-stability. The following example on S^2 indicates this subtle point. See Figure 5.4.

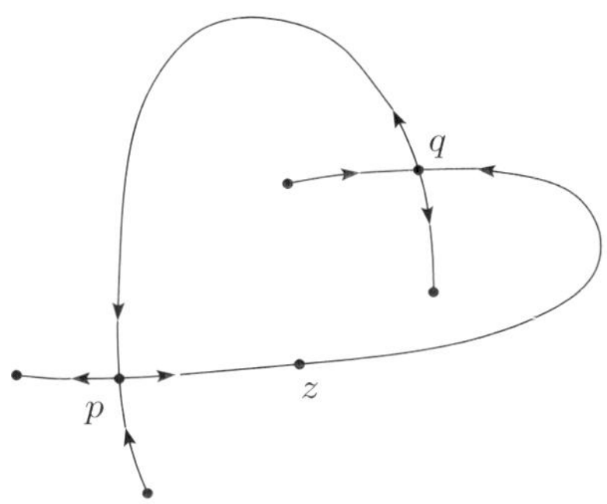

Figure 5.4. The Ω-explosion. z is wandering because a neighborhood of it splits into three parts such that the inside part goes to the inside sink, the outside part goes to the outside sink, and the interval part on the cycle goes to the saddle.

Denote by f the diffeomorphism in Figure 5.4. It is straightforward to check by definition that every point z on the "cycle" connecting the two saddles p and q is wandering, except p and q themselves. (Note that this is different from Figure 5.2 where, by the λ-lemma, every point on the homoclinic loop is nonwandering.) Also, every point not on the cycle goes either positively to a sink or negatively to a source. Thus the nonwandering set $\Omega(f)$ consists of the six hyperbolic fixed points only. Hence f satisfies Axiom A.

But f is not Ω-stable. With an arbitrarily small C^r perturbation near z, one can make the stable and unstable manifolds of the new map g transverse at z and hence, by the λ-lemma, make every point of the other saddle connection nonwandering, yielding an "*Ω-explosion*". Since $\Omega(f)$ has six points but $\Omega(g)$ is infinite, f is not Ω-stable.

We need to pay some attention here to the construction of the perturbation near z. Let h be a small rotation supported on a small neighborhood $B(z,\delta)$ of z, meaning h is a rotation on $B(z,\delta/3)$ and is the identity outside $B(z,\delta)$. The number $\delta > 0$ should be small enough so that $f^n(B(z,\delta)) \cap B(z,\delta) = \emptyset$ for all $n \neq 0$ (this is possible since z is wandering) and so that the saddle connection from p to q cuts through $B(z,\delta)$ only once and so that, finally, the other saddle connection from q to p does not intersect $B(z,\delta)$.

Take our perturbation to be $g = h \circ f$. Then the saddle connection from q to p remains unchanged since it is outside of the support of the perturbation. We verify that $W^s(z,g)$ and $W^u(z,g)$ intersect transversely at z. Take $r > 0$ sufficiently small so that $f(W^s_r(z,f))$ is disjoint from $B(z,\delta)$ and hence is not affected by h and hence will not be affected by h for further positive iterates. Thus $W^s_r(z,g) = W^s_r(z,f)$. In contrast, for $g^{-1} = f^{-1} \circ h^{-1}$, it is $h(W^u_r(z,f))$ that is rotated by h^{-1} to get onto $W^u_r(z,f)$ and then mapped by f^{-1} to get disjoint from $B(z,\delta)$ and then will not be affected by h for further negative iterates. Thus $W^u_r(z,g) = h(W^s_r(z,f))$. In other words, exactly one (but not both) of the two old local manifolds rotates. Thus the two new local manifolds do intersect transversely, yielding an Ω-explosion.

In general, it is not obvious if an Ω-explosion must contradict the Ω-stability (though it certainly contradicts the ϵ-Ω-stability) if $\Omega(f)$ and $\Omega(g)$ are both infinite. The answer turns out to be yes because Ω-stability turns out to be equivalent to ϵ-Ω-stability, at least for the C^1 topology. We will come back to this striking point in Section 6.5.

Another interesting point indicated by Figure 5.4 is that, though a basic set is always contained in a chain class, it may not equal the chain class, even if f satisfies Axiom A. The two saddles p and q in Figure 5.4 are each a basic set but are contained in the same chain class. Thus whether a basic set B_i is a chain class is not a local problem determined by information in a neighborhood of B_i. Corollary 5.12 below specifies a situation when a basic set is a chain class.

5.3. No-cycle and Ω-stability

The existence of the "cycle" in the example of Figure 5.4 is the key to the gap between Axiom A and Ω-stability. We define cycles generally.

Let X be a compact metric space, and let $f : X \to X$ be a homeomorphism. Let $\Lambda_1, \ldots, \Lambda_k$ be finitely many disjoint compact invariant sets of f. Write $\Lambda = \Lambda_1 \cup \cdots \cup \Lambda_k$. Define $\Lambda_i \rightharpoonup \Lambda_j$ if

$$W^u(\Lambda_i) \cap W^s(\Lambda_j) - \Lambda \neq \emptyset.$$

Briefly, $\Lambda_i \rightharpoonup \Lambda_j$ means there is $x \in X$ *outside* Λ that goes from Λ_i to Λ_j. Note that when $i \neq j$, $W^u(\Lambda_i) \cap W^s(\Lambda_j) \neq \emptyset$ implies $\Lambda_i \rightharpoonup \Lambda_j$ automatically, but not when $i = j$.

It is straightforward to check that the binary relation $\rightharpoonup$ is not reflexive, nor symmetric, nor transitive. We say $\Lambda_{i_1}, \ldots, \Lambda_{i_m}$ form a *cycle* of $\{\Lambda_i\}$ if

$$\Lambda_{i_1} \rightharpoonup \Lambda_{i_2} \rightharpoonup \cdots \rightharpoonup \Lambda_{i_m} \rightharpoonup \Lambda_{i_1}.$$

We say the $\{\Lambda_i\}$ satisfy the *no-cycles condition*, or simply *have no cycles*, if no subset of $\{\Lambda_i\}$ forms a cycle. The next theorem is due to Pugh-Shub (1970).

Theorem 5.6. *Let $f : X \to X$ be a homeomorphism, and let $\Omega(f) = \Lambda_1 \cup \cdots \cup \Lambda_k$ be a decomposition into finitely many disjoint compact invariant sets. If the $\{\Lambda_i\}$ have no cycles, then for any neighborhood V of $\Omega(f)$ in M, there is a C^0 neighborhood $\mathcal{U}$ of f such that $\Omega(g) \subset V$ for every $g \in \mathcal{U}$.*

Briefly, if there is no Ω-cycle, then there is no Ω-explosion.

Proof. Suppose to the contrary that there are a compact neighborhood V_0 of $\Omega(f)$ in X and a sequence $g_n \to f$ in the C^0 topology, together with $a_n \in \Omega(g_n)$ such that $a_n \notin V_0$. We may assume $a_n \to b_1 \notin \Omega(f)$. By the definition of nonwandering point, there are sequences $x_n \to a_n$ and $k_n \to \infty$ such that $y_n = g_n^{k_n}(x_n) \to a_n$. Hence $x_n \to b_1$, $y_n \to b_1$.

We single out a repeatedly used argument, namely the following lemma. Denote the g_n-orbit from x_n to y_n simply by $[x_n, y_n]$, and the g_n-orbit from $g_n(x_n)$ to $g_n^{-1}(y_n)$ by (x_n, y_n). We may write something like $[x_n, y_n]_{g_n}$. But since $[x_n, y_n]$ always denotes this specific piece of the g_n-orbit (from x_n to y_n), for simplicity we have omitted the subscript g_n. We remark that by $W^s(\Lambda_i)$ below we will mean $W^s(\Lambda_i, f)$. Here is the lemma.

Lemma 5.7. *Let $x, y \notin \Omega(f)$. Let $[x_n, y_n]$ be an orbit-arc of g_n such that $x_n \to x$, $y_n \to y$. Assume $x \in W^s(\Lambda_i)$ for some i. Then there are $z \in W^u(\Lambda_i) - \Omega(f)$ and $z_n \in (x_n, y_n)$ for large n such that $z_n \to z$.*

Proof. Take a compact neighborhood U of Λ_i such that

$$x, y \notin U, \quad U \cap \Lambda_j = \emptyset, \quad f(U) \cap \Lambda_j = \emptyset, \quad \forall\, j \neq i.$$

Let $p_n \in [x_n, y_n]$ be the point in $[x_n, y_n]$ that is closest to Λ_i. Since $x \in W^s(\Lambda_i)$ and $x_n \to x$ and $g_n \to f$, taking subsequences if necessary, we may assume

$$d(p_n, \Lambda_i) \to 0, \quad n \to \infty.$$

In particular, $p_n \in (x_n, y_n)$ for large n. Since $y_n \to y \notin U$, there is $m_n \geq 1$ for large n such that

$$p_n,\ g_n(p_n),\ \ldots,\ g_n^{m_n - 1}(p_n) \in \operatorname{int} U,$$

but

$$g_n^{m_n}(p_n) \notin \operatorname{int} U.$$

Then

$$z_n = g_n^{m_n}(p_n) \in g_n(\operatorname{int} U) - \operatorname{int} U \subset g_n(U) - \operatorname{int} U.$$

Note that $z_n \in (x_n, y_n)$. See Figure 5.5. Since $p_n \to \Lambda_i$, we have

$$m_n \to +\infty.$$

Also, since $g_n \to f$, if n is large, then

$$g_n(U) \cap \Lambda_j = \emptyset, \quad \forall\, j \neq i.$$

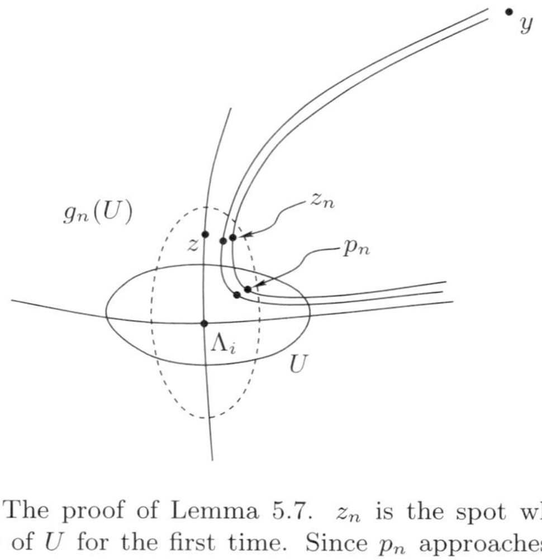

Figure 5.5. The proof of Lemma 5.7. z_n is the spot where p_n gets positively out of U for the first time. Since p_n approaches Λ_i, z_n will remain negatively in U for a finite but longer and longer time. Passing to the limit, z will have its whole negative orbit inside U.

Let z be a limit point of $\{z_n\}$. Then $z \notin \Omega(f)$. It remains to prove $z \in W^u(\Lambda_i)$. Since the first m_n iterates of z_n with respect to g_n^{-1} are contained in U and since $g_n \to f$ and $m_n \to \infty$, it follows that $f^{-k}z \in U$ for all $k \geq 1$. Hence $\alpha(z) \subset \Lambda_i$, meaning $z \in W^u(\Lambda_i)$. This proves Lemma 5.7. □

Now we finish the proof of Theorem 5.6. By Theorem 5.4, there is i_1 such that $b_1 \in W^s(\Lambda_{i_1})$. By Lemma 5.7, there are $b_2 \in W^u(\Lambda_{i_1}) - \Omega(f)$ and $z_n \in (x_n, y_n)$ for large n such that $z_n \to b_2$. By Theorem 5.4, there is i_2 such that $b_2 \in W^s(\Lambda_{i_2})$. By Lemma 5.7, there are $b_3 \in W^u(\Lambda_{i_2}) - \Omega(f)$ and $z_n' \in (z_n, y_n)$ for large n such that $z_n' \to b_3$, and so on. Since there are only finitely many Λ_i, this will trace out a cycle, a contradiction. This proves Theorem 5.6. □

A diffeomorphism $f : M \to M$ with $\overline{\mathrm{P}(f)}$ hyperbolic is said to *satisfy the no-cycle condition* if the basic sets of $\overline{\mathrm{P}(f)}$ have no cycles. Here is the celebrated Ω-stability theorem of Smale (1970).

Theorem 5.8 (The Ω-stability theorem). *Let $f : M \to M$ be a diffeomorphism. If f satisfies Axiom A and the no-cycle condition, then f is C^r Ω-stable for any $r \geq 1$.*

Proof. It suffices to prove f is C^1 Ω-stable. In fact we prove f is C^1 ϵ-Ω-stable.

By Theorem 5.3, $\Omega(f)$ is isolated. Let U be an isolating neighborhood of $\Omega(f)$ in M. Let $\epsilon > 0$ be given. By Theorem 4.23, there is a C^1 neighborhood $\mathcal{U}$ of f such that, for any $g \in \mathcal{U}$, the maximal invariant set Γ_g of g in U is isolated in U and there is an (onto) homeomorphism

$$h : \Omega(f) \to \Gamma_g$$

such that

$$hf = gh, \quad d(h, id) \leq \epsilon.$$

Let $\{B_i\}$ be the basic sets of f. Since the $\{B_i\}$ have no cycles, by Theorem 5.6, there is a C^1 neighborhood $\mathcal{U}_1 \subset \mathcal{U}$ of f such that $\Omega(g) \subset U$ for every $g \in \mathcal{U}_1$. It remains to prove

$$\Omega(g) = \Gamma_g.$$

Since Γ_g is the maximal invariant set of g in U and $\Omega(g) \subset U$, the "⊂" part is obvious. The "⊃" part is given by

$$\begin{aligned} \Gamma_g &= h(\Omega(f)) = h(\overline{\mathrm{P}(f)}) \\ &= \overline{h(\mathrm{P}(f))} \subset \overline{\mathrm{P}(g)} \subset \Omega(g). \end{aligned}$$

This proves Theorem 5.8. □

The original proof of Smale for Theorem 5.8 uses filtration (see the next section for the definition). The proof given here using Theorem 5.6 is taken from Pugh-Shub (1970).

Note that to prove the "$\supset$" part we cannot argue like this: h is a conjugacy, and "a conjugacy preserves the nonwandering set"; hence $h(\Omega(f)) \subset \Omega(g)$. This is because h is defined merely on $\Omega(f)$; hence the "nonwandering set" that h preserves is $\Omega(f|_{\Omega(f)})$, which is generally different from $\Omega(f)$.

5.4. Equivalent descriptions

The Axiom A plus no-cycle condition is about the nonwandering set $\Omega(f)$. Newhouse (1972, 1980) observed that the condition can be formulated in an equivalent way using the limit set $\mathrm{L}(f)$, which is smaller than $\Omega(f)$. Franke-Selgrade (1977) found another equivalent description using the chain recurrent set $\mathrm{CR}(f)$, which is larger than $\Omega(f)$. In this section we introduce their results with somewhat different treatment.

Theorem 5.9. *Let $f : X \to X$ be a homeomorphism, and let $\mathrm{L}(f) = \Lambda_1 \cup \cdots \cup \Lambda_k$ be a decomposition into finitely many disjoint compact invariant sets. If $\{\Lambda_i\}$ have no cycles, then $\mathrm{L}(f) = \mathrm{CR}(f)$.*

Proof. First we single out a repeatedly used argument below, namely the following lemma.

Lemma 5.10. *Let $x \in \mathrm{CR}(f) - \mathrm{L}(f)$. Assume $x \in W^s(\Lambda_i)$ for some i. Then there is $z \in W^u(\Lambda_i)$ such that $z \in \mathrm{CR}(f) - \mathrm{L}(f)$.*

Proof. The proof is similar to Lemma 5.7. Take a compact neighborhood U of Λ_i such that

$$x \notin U, \quad U \cap \Lambda_j = \emptyset, \quad f(U) \cap \Lambda_j = \emptyset, \quad \forall\, j \neq i.$$

For every $n \geq 1$, there is a periodic $1/n$-chain

$$C_n = \{x_0^n,\ x_1^n,\ \ldots,\ x_{j_n}^n\}$$

such that $x_0^n \to x$. Let $x_{\alpha_n}^n$ be the point in C_n that is closest to Λ_i. Since $x \in W^s(\Lambda_i)$, taking subsequences if necessary, we may assume

$$d(x_{\alpha_n}^n, \Lambda_i) \to 0, \quad n \to \infty.$$

Then there is $m_n \geq 1$ for large n such that

$$x_{\alpha_n}^n,\ x_{\alpha_n+1}^n,\ \ldots,\ x_{\alpha_n+m_n-1}^n \in \operatorname{int} U,$$

but

$$x_{\alpha_n+m_n}^n \notin \operatorname{int} U.$$

Then

$$z^n = x_{\alpha_n+m_n}^n \in B(f(U), 1/n) - \operatorname{int} U.$$

Since $x^n_{\alpha_n} \to \Lambda_i$, we have

$$m_n \to +\infty.$$

Also, if n is large, then

$$B(f(U), 1/n) \cap \Lambda_j = \emptyset, \quad \forall\, j \neq i.$$

Let z be a limit point of $\{z^n\}$. Then $z \in \mathrm{CR}(f) - \mathrm{L}(f)$. It remains to prove $z \in W^u(\Lambda_i)$. Since the first m_n points in the periodic $1/n$-chain C_n starting with $z^n = x^n_{\alpha_n+m_n-1}$ counting backwards are contained in U and since $z^n \to z$ and $m_n \to \infty$, it follows that $f^{-k}z \in U$ for all $k \geq 1$. Hence $\alpha(z) \subset \Lambda_i$, meaning $z \in W^u(\Lambda_i)$. This proves Lemma 5.10. ☐

Now we finish the proof of Theorem 5.9. Suppose to the contrary that there is $b_1 \in \mathrm{CR}(f) - \mathrm{L}(f)$. By Theorem 5.4, there is i_1 such that $b_1 \in W^s(\Lambda_{i_1})$. By Lemma 5.10, there is $b_2 \in W^u(\Lambda_{i_1})$ such that $b_2 \in \mathrm{CR}(f) - \mathrm{L}(f)$. By Theorem 5.4, there is i_2 such that $b_2 \in W^s(\Lambda_{i_2})$. By Lemma 5.10, there is $b_3 \in W^u(\Lambda_{i_2})$ such that $b_3 \in \mathrm{CR}(f) - \mathrm{L}(f)$, and so on. Since there are only finitely many Λ_i, this will trace out a cycle, a contradiction. This proves Theorem 5.9. ☐

Recall that we say a diffeomorphism $f : M \to M$ with $\overline{\mathrm{P}(f)}$ hyperbolic satisfies the no-cycle condition if the basic sets of $\overline{\mathrm{P}(f)}$ have no cycles.

Theorem 5.11. *Let $f : M \to M$ be a diffeomorphism. The following three conditions are equivalent:*

(1) *f satisfies Axiom A and the no-cycle condition.*

(2) *$\mathrm{L}(f)$ is hyperbolic and satisfies the no-cycle condition.*

(3) *$\mathrm{CR}(f)$ is hyperbolic.*

Proof. (1) $\Rightarrow$ (2): Immediate from the definitions.

(2) $\Rightarrow$ (3): By Theorem 4.31, $\mathrm{L}(f) = \overline{\mathrm{P}(f)}$. By Theorem 5.2, $\mathrm{L}(f)$ decomposes into the basic sets B_i of $\overline{\mathrm{P}(f)}$. Since the $\{B_i\}$ have no cycles, by Theorem 5.9, $\mathrm{L}(f) = \mathrm{CR}(f)$.

(3) $\Rightarrow$ (1): By Theorem 4.32, $\mathrm{CR}(f) = \overline{\mathrm{P}(f)}$. Then f satisfies Axiom A. Also, $\mathrm{CR}(f)$ decomposes into the basic sets B_i of $\overline{\mathrm{P}(f)}$. It remains to prove that the $\{B_i\}$ have no cycles. Suppose there is a cycle

$$z_1 \in W^u(B_{i_1}) \cap W^s(B_{i_2}) - \mathrm{CR}(f),\ z_2 \in W^u(B_{i_2}) \cap W^s(B_{i_3}) - \mathrm{CR}(f),$$
$$\ldots, z_m \in W^u(B_{i_m}) \cap W^s(B_{i_1}) - \mathrm{CR}(f).$$

We emphasize that, by the definition of $\mathrm{CR}(f)$-cycle, $z_i \notin \mathrm{CR}(f)$. But since each B_i is transitive, every z_i is forced to be chain recurrent, a contradiction. This proves Theorem 5.11. ☐

Remark. (1) $\Leftrightarrow$ (2) is due to Newhouse (1972, 1980). Examples show that $\mathrm{L}(f)$ being hyperbolic is strictly weaker than $\Omega(f)$ being hyperbolic, let alone Axiom A. Thus the no-cycle condition is crucial here to make (1) and (2) equivalent. (1) $\Leftrightarrow$ (3) is due to Franke-Selgrade (1977).

Corollary 5.12. *If f satisfies Axiom A plus the no-cycle condition, then*

$$\overline{\mathrm{P}(f)} = \mathrm{L}(f) = \Omega(f) = \mathrm{CR}(f),$$

and every basic set B_i of $\overline{\mathrm{P}(f)}$ is a chain class.

Proof. Let f satisfy Axiom A plus the no-cycle condition. By Theorem 5.11, $\mathrm{CR}(f)$ is hyperbolic. By Theorem 4.32, $\mathrm{CR}(f) = \overline{\mathrm{P}(f)}$.

Let the $\{B_i\}$ be the basic sets of $\overline{\mathrm{P}(f)}$. We prove B_1, say, is a chain class. Being transitive, B_1 is contained in a chain class C. Now

$$C \subset \mathrm{CR}(f) = B_1 \cup \cdots \cup B_k.$$

By Theorem 1.7, C is indecomposable and hence is contained in one of the B_i, which can only be B_1. Thus $B_1 = C$, proving Corollary 5.12. $\square$

Another striking feature for Axiom A plus no-cycle systems is the existence of filtration. Let $f : X \to X$ be a homeomorphism. By a *filtration* of f we mean finitely many compact sets

$$\emptyset = X_0 \subset X_1 \subset \cdots \subset X_k = X$$

such that

$$f(X_i) \subset \operatorname{int} X_i, \quad i = 0, \ldots, k.$$

Thus a filtration is just a nested sequence of (compact) trapping regions. See Figure 5.6.

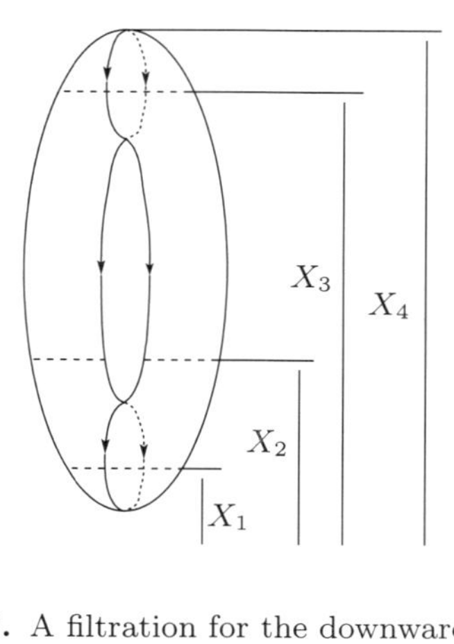

Figure 5.6. A filtration for the downward map on $\mathbb{T}^2$.

The next proposition states some obvious properties of filtration.

Proposition 5.13. *Let $f : X \to X$ be a homeomorphism. If $\{X_i\}_{i=0}^k$ is a filtration of f, then the maximal invariant set Γ_i of f in $X_i - X_{i-1}$ is isolated with isolating neighborhood $X_i - X_{i-1}$, and the $\{\Gamma_i\}$ have no cycles. Moreover, $\Gamma_1 \cup \cdots \cup \Gamma_k \supset \mathrm{CR}(f)$.*

Proof. This proposition follows directly from a simple fact that since X_i is a trapping region, if $x \notin X_i$ but $fx \in X_i$, then $x \notin \mathrm{CR}(f)$. See the first half of the proof of Lemma 1.15. We stop the proof here. □

Theorem 5.14. *Let $f : X \to X$ be a homeomorphism. If $\mathrm{CR}(f)$ has only finitely many chain classes $C_1, \ldots, C_k$, then rearranging the subscripts if necessary, there is a filtration $\{X_i\}_{i=0}^k$ such that the maximal invariant set of f in $X_i - X_{i-1}$ is exactly C_i.*

Proof. We use Theorem 1.13, the fundamental theorem of dynamical systems of Conley. See Shub (1987) for a nice treatment of filtration theory without using Conley theory.

Let $\phi : X \to \mathbb{R}$ be a Lyapunov function of f. Rearranging the subscripts if necessary, we may assume

$$\phi(C_1) < \phi(C_2) < \cdots < \phi(C_k).$$

Insert numbers $a_i \in \mathbb{R}$ such that

$$\phi(C_1) < a_1 < \phi(C_2) < a_2 < \cdots < \phi(C_k) < a_k.$$

Let

$$X_i = \phi^{-1}(-\infty, a_i].$$

Then $\{X_i\}$ is a filtration such that

$$C_i \subset X_i - X_{i-1}.$$

We prove that the maximal invariant set Γ_i of f in $X_i - X_{i-1}$ is exactly C_i. It suffices to prove $\Gamma_i \subset C_i$.

Let $x \in \Gamma_i$. Then

$$\omega(x) \subset \mathrm{CR}(f) \cap \Gamma_i \subset C_i.$$

Likewise, $\alpha(x) \subset C_i$. Then $x \in C_i$ since C_i is a chain class. This proves Theorem 5.14. □

Theorem 5.15. *If f satisfies Axiom A and the no-cycle condition, then rearranging the subscripts if necessary, there is a filtration $\{M_i\}$ such that the maximal invariant set of f in $M_i - M_{i-1}$ is exactly the basic set B_i.*

Proof. The proof is immediate from Corollary 5.12 and Theorem 5.14. □

We have investigated the structural stability for the nonwandering set $\Omega(f)$. At the end of this chapter we state without proof the structural stability result for the whole manifold. The statement uses a condition called strong transversality, which we now define.

By Theorem 5.5, if $\mathrm{L}(f)$ is hyperbolic, then every point $x \in M$, including the wandering ones, shares the same stable manifold with a point in $\mathrm{L}(f)$; hence $W^s(x)$ is an immersed submanifold of M. Likewise for $W^u(x)$. With this in mind we give the following definition.

A diffeomorphism $f : M \to M$ with $\mathrm{L}(f)$ hyperbolic is said to *satisfy the strong transversality condition* if for every $x \in M$, $W^s(x)$ and $W^u(x)$ are transverse at x.

Note that if $x \in \mathrm{L}(f)$, by the stable manifolds theorem of a hyperbolic set, $W^s(x)$ and $W^u(x)$ are transverse at x already (with complementary dimensions). Thus the strong transversality condition is actually a condition put on $x \notin \mathrm{L}(f)$, and in this case it allows $W^s(x)$ and $W^u(x)$ to have extra dimensions. The simplest example could be the north-south poles map f on S^2, for which $\mathrm{L}(f)$ consists of the two (hyperbolic) poles. At every $x \notin \mathrm{L}(f)$, $W^s(x)$ and $W^u(x)$ are transverse at x with extra dimensions.

We state without proof the celebrated structural stability theorem. Recall that f is C^r *structurally stable* if there is a C^r neighborhood $\mathcal{U}$ of f in $\mathrm{Diff}^r(M)$ such that every $g \in \mathcal{U}$ is topologically conjugate to f. After a number of previous works by Palis (1968), Palis-Smale (1970), Robbin (1971), and de Melo (1973), Robinson (1976) eventually solved this remarkable problem:

Theorem 5.16 (The structural stability theorem)**.** *Let $f : M \to M$ be a diffeomorphism. If f satisfies Axiom A and the strong transversality condition, then f is C^r structurally stable for any $r \geq 1$.*

Exercises

Exercise 5.1. Find a hyperbolic set Λ with

$$W^s(\Lambda) \neq \bigcup_{x \in \Lambda} W^s(x).$$

Exercise 5.2. Let p and q be two hyperbolic periodic points of f. We say p and q are *homoclinically related* if $W^s(\mathrm{Orb}(p))$ and $W^u(\mathrm{Orb}(q))$ have a transverse intersection and $W^s(\mathrm{Orb}(q))$ and $W^u(\mathrm{Orb}(p))$ have a transverse intersection. The closure of the set of hyperbolic periodic points that are homoclinically related to p is called the *homoclinic class* of p, denoted $H(p, f)$.

Prove the following:

(1) $H(p, f)$ is transitive.

(2) If $H(p, f)$ is not a single (periodic) orbit, it coincides with the closure of the set of transverse homoclinic points of p.

Exercise 5.3. Show that every basic set of f (with $\overline{\mathrm{P}(f)}$ hyperbolic) is a homoclinic class.

Exercise 5.4. Let $\Omega_i = \overline{P}_{i_1} \cup \cdots \cup \overline{P}_{i_r}$ be the cyclic union described in the proof of the spectral decomposition theorem. Show that f^r restricted to each $\overline{P}_{i_j}$ is topologically mixing.

Exercise 5.5. Assume Λ is a basic set of f (with $\overline{\mathrm{P}(f)}$ hyperbolic). Let p be a periodic point in Λ. Prove $W^s(\mathrm{Orb}(p))$ is dense in $W^s(\Lambda)$. Similarly, prove $W^u(\mathrm{Orb}(p))$ is dense in $W^u(\Lambda)$.

Exercise 5.6. In the global horseshoe map $f : S^2 \to S^2$ one may arrange that the nonwandering set of f is the Cantor set Λ plus a hyperbolic source in the lower hemisphere and a hyperbolic sink in the upper hemisphere. Verify that f satisfies Axiom A and the no-cycle condition, and illustrate a filtration for f.

A compact invariant set $\Lambda \subset M$ of f is called *Lyapunov stable* if for any neighborhood U of Λ, there exists a neighborhood $V \subset U$ of Λ such that for any $x \in V$ and any $n \geq 0$, $f^n(x) \in U$.

Exercise 5.7. Prove that any attracting set (see Section 1.4) is Lyapunov stable, but not vice versa.

Exercise 5.8. Find a compact invariant set Λ with a compact neighborhood U such that $\omega(x) \subset \Lambda$ for any $x \in U$, but Λ is not attracting.

Exercise 5.9. Let Λ be Lyapunov stable. Prove $W^u(x) \subset \Lambda$ for any $x \in \Lambda$.

Exercise 5.10. Let Λ be Lyapunov stable. Assume there exists a neighborhood U of Λ such that for any $x \in U$, $\omega(x) \cap \Lambda \neq \emptyset$. Prove Λ is attracting.

Exercise 5.11. Let Λ be Lyapunov stable. Assume Λ is hyperbolic. Prove Λ is attracting.

Exercise 5.12. Let $p \in M$ be a hyperbolic periodic point of f. Prove that if $x \in \overline{W^s(\mathrm{Orb}(p))} \cap \overline{W^u(\mathrm{Orb}(p))}$, then $x \in \Omega(f)$. Moreover, if p and q are two hyperbolic periodic points that are homoclinically related and if $x \in \overline{W^s(\mathrm{Orb}(p))} \cap \overline{W^u(\mathrm{Orb}(q))}$, then $x \in \Omega(f)$.

Exercise 5.13. Let $p \in M$ be a hyperbolic periodic point of f. Prove that if $\overline{W^s(\mathrm{Orb}(p))} = \overline{W^u(\mathrm{Orb}(p))} = M$, then f is transitive. If, in addition, $p \in M$ is a fixed point, then f is mixing.

Exercise 5.14. Let $F : \mathbb{T}^2 \times \mathbb{T}^2 \to \mathbb{T}^2 \times \mathbb{T}^2$ be a diffeomorphism of the form of skew product $F(x, y) = (f(x), g(x, y))$ ("fiber preserving"). Denote simply by $0 \in \mathbb{T}^2$ the point $\pi(0)$, where $\pi : \mathbb{R}^2 \to \mathbb{T}^2$ is the covering map defined in Section 3.3. Prove that if $f : \mathbb{T}^2 \to \mathbb{T}^2$ and $g_0 = g(0, \cdot) : \mathbb{T}^2 \to \mathbb{T}^2$ are Anosov toral automorphisms, respectively, then F is mixing.

Exercise 5.15. Find a diffeomorphism f with infinitely many chain classes.

Exercise 5.16. Prove that there can be no cycle among any (finite) number of chain classes.

Exercise 5.17. Prove that if a chain class is hyperbolic, then it is isolated, transitive, and with periodic points dense.

A hyperbolic set $\Lambda \subset M$ is said to have *homogeneous index* if, for the hyperbolic splitting $E^s(x) \oplus E^u(x)$, $\dim E^s(x)$ is constant for any $x \in \Lambda$. The constant is called the *index* of Λ.

Exercise 5.18. Let $y \in M$. If $\omega(y)$ is hyperbolic, prove $\omega(y)$ has homogeneous index.

Exercise 5.19. Let $\Lambda_1 \subset M$ and $\Lambda_2 \subset M$ be two transitive hyperbolic sets of different indices. Assume there is a cycle between Λ_1 and Λ_2. Prove $\mathrm{CR}(f)$ is not hyperbolic.

Exercise 5.20. A point $x \in M$ is called a *homoclinic tangency* of a hyperbolic periodic point $p \in M$ if $W^s(\mathrm{Orb}(p))$ and $W^u(\mathrm{Orb}(p))$ intersect nontransversely at x. Show that if f admits a homoclinic tangency, then $\Omega(f)$ is not hyperbolic.

Exercise 5.21. Prove there is no 1-cycle among the basic sets of any Axiom A diffeomorphism f; that is, for any basic set Λ of f, $W^u(\Lambda) \cap W^s(\Lambda) = \Lambda$.

Exercise 5.22. Let $f : M \to M$ be Anosov. If $\mathrm{CR}(f) = M$, prove that f is transitive (we always assume M is connected).

Exercise 5.23. Let $f : M \to M$ be Anosov. Prove f satisfies Axiom A.

Exercise 5.24. Let $f : M \to M$ satisfy Axiom A. Prove that if f satisfies the strong transversality, then f satisfies the no-cycle condition.

Chapter 6

Quasi-hyperbolicity and linear transversality

In this chapter we give a short account for the theory of quasi-hyperbolicity and linear transversality. The two notions are in a sense dual to each other. They do not assume uniform contraction or expansion but merely some limit properties. Nevertheless by compactness of the set they turn out to guarantee uniform hyperbolicity on the chain recurrent part of the set. In particular, they give equivalent descriptions for compact hyperbolic sets that look much weaker. This provides alternate angles for looking at hyperbolicity, the main concept of the previous chapters.

The theory is based on works of several authors. Sacker-Sell (1974), Selgrade (1975), and Churchill-Franke-Selgrade (1977) investigated quasi-hyperbolicity. Mañé (1977a, 1977b) and Liao (1980b) investigated quasi-hyperbolicity and linear transversality.

6.1. The simplest setting

Let E be a Euclidean space. A linear isomorphism $A : E \to E$ is called *quasi-hyperbolic* if every nonzero vector has unbounded orbit, that is, if

$$B^s \cap B^u = \{0\},$$

where

$$B^s = \{v \in E \mid \{|A^n(v)|\}_{n=0}^{\infty} \text{ is bounded}\},$$
$$B^u = \{v \in E \mid \{|A^{-n}(v)|\}_{n=0}^{\infty} \text{ is bounded}\}.$$

We say A satisfies the *linear transversality condition* if

$$D^s + D^u = E,$$

where

$$D^s = \{v \in E \mid |A^n(v)| \to 0,\ n \to \infty\},$$
$$D^u = \{v \in E \mid |A^{-n}(v)| \to 0,\ n \to \infty\}.$$

Clearly, B^s, B^u, D^s, and D^u are linear subspaces of E, invariant under A.

Theorem 6.1. *Let $A : E \to E$ be a linear isomorphism. The following conditions are equivalent:*

(1) *A is hyperbolic.*

(2) *A is quasi-hyperbolic.*

(3) *A satisfies the linear transversality condition.*

Proof. Let A be hyperbolic. By Theorem 2.2, $E^s = B^s = D^s$, and $E^u = B^u = D^u$. Then (1) $\Rightarrow$ (2) and (1) $\Rightarrow$ (3).

We prove (2) $\Rightarrow$ (1). Suppose A is not hyperbolic. Then A has an eigenvalue λ of absolute value 1. If λ is real, then any eigenvector v has $|A^n v| = |v|$ for all $n \in \mathbb{Z}$. If λ is not real, then A has an invariant plane P such that $A|_P$ is conjugate linearly to a rotation. In both cases, there is $v \neq 0$ whose orbit is bounded. Hence A is not quasi-hyperbolic. This proves (2) $\Rightarrow$ (1).

Finally we prove (3) $\Rightarrow$ (1). The same consideration on eigenvalues shows that all eigenvalues of $A|_{D^s}$ are less than 1 in absolute value. Hence $A|_{D^s}$ is contracting. Likewise $A|_{D^u}$ is expanding. In particular, $D^s \cap D^u = \{0\}$. Now $D^s + D^u = E$. Hence $D^s \oplus D^u = E$. Then A is hyperbolic. This proves (3) $\Rightarrow$ (1). □

Thus, in this simplest setting, quasi-hyperbolicity and linear transversality turn out to be equivalent to hyperbolicity, though looking much weaker by the definitions.

6.2. Quasi-hyperbolicity

Quasi-hyperbolicity and linear transversality can be defined, in a similar way, for any compact invariant set $\Lambda \subset M$ of f. Then the two notions will no longer be equivalent to hyperbolicity but will still capture hyperbolicity to a great extent. In this chapter we explore this. The approach is taken from Mañé (1977a, 1977b). Liao (1980b) developed the same theory through a different approach called the obstruction set.

This section contains the heart of the theory. First we give an equivalent condition for contraction.

Theorem 6.2. *Let $\Lambda \subset M$ be a compact invariant set of f, and let E be a Tf-invariant subbundle of $T_\Lambda M$. The following two conditions are equivalent:*

(1) *There are $0 < \lambda < 1$ and $C \geq 1$ such that $|Tf^n(v)| \leq C\lambda^n|v|$, $\forall v \in E, n \geq 1$.*

(2) *For any $0 \neq v \in E$, there is $m = m(v)$ such that $|Tf^m(v)| < |v|$.*

Proof. Condition (1) is what we usually mean by a *contraction.* Clearly (1) $\Rightarrow$ (2). We prove (2) $\Rightarrow$ (1). Since Λ is compact, so is the unit sphere bundle $E(1) = \{v \in E \mid |v| = 1\}$. Then there are a positive integer N and a number $0 < \mu < 1$ such that for any $v \in E$, there is $0 \leq m \leq N$ such that

$$|Tf^m(v)| \leq \mu|v|.$$

Set

$$\lambda = \mu^{1/N}, \ C = \lambda^{-N} \cdot \max\{|Tf^n(v)| \mid v \in E, |v| = 1, 0 \leq n \leq N\}.$$

Let $v \in E$ and $n \geq 0$ be given. There is $0 \leq m_1 \leq N$ for v such that $|Tf^{m_1}(v)| \leq \mu|v|$. Likewise, there is $0 \leq m_2 \leq N$ for $Tf^{m_1}(v)$ such that $|Tf^{m_1+m_2}(v)| \leq \mu|Tf^{m_1}(v)| \leq \mu^2|v|$. This gives rise to a positive number k such that

$$m_1 + m_2 + \cdots + m_k \leq n \leq m_1 + m_2 + \cdots + m_k + N.$$

Then

$$|Tf^n(v)| \leq C\lambda^N \cdot \mu^k|v| = C\lambda^N \cdot \lambda^{kN}|v| \leq C\lambda^N \cdot \lambda^{m_1+\cdots+m_k}|v| \leq C\lambda^n|v|.$$

This proves Theorem 6.2. $\square$

For $x \in M$, denote

$$B^s(x) = \{v \in T_xM \mid \{|Tf^n(v)|\}_{n=0}^\infty \text{ is bounded}\},$$
$$B^u(x) = \{v \in T_xM \mid \{|Tf^{-n}(v)|\}_{n=0}^\infty \text{ is bounded}\}.$$

These are linear subspaces of T_xM, invariant under Tf, as family.

A compact invariant set Λ of f is called *quasi-hyperbolic* if every nonzero vector of $T_\Lambda M$ has unbounded orbit, that is, if $B^s(x) \cap B^u(x) = \{0\}$ for all $x \in \Lambda$.

Remark. We emphasize that compactness is assumed in the definition of quasi-hyperbolic set. Without compactness the results about quasi-hyperbolic sets below will not be true. This is like the situation of Theorem 6.2 where the assumption of compactness of Λ is necessary and crucial. *For this reason invariant sets Λ considered in this chapter are all assumed to be compact.* Note that this is contrary to the definition of hyperbolic set, where compactness is automatic since, by Theorem 4.3, the uniform (C, λ)-rates of contraction and expansion pass to the closure.

Lemma 6.3. *Λ is quasi-hyperbolic for f if and only if there is a positive integer N such that for any $0 \neq v \in T_\Lambda M$, there is $-N \leq m \leq N$ such that $|Tf^m(v)| > 2|v|$.*

Proof. Let Λ be quasi-hyperbolic for f. For any $v \in T_\Lambda M$ with $|v| = 1$, there is an integer $m = m(v)$, positive or negative, such that $|Tf^m(v)| > 2$. By the compactness of the unit sphere bundle of $T_\Lambda M$, there is a positive integer N such that for any $0 \neq v \in T_\Lambda M$, there is $-N \leq m \leq N$ such that $|Tf^m(v)| > 2|v|$. Conversely, assume for any $0 \neq v \in T_\Lambda M$, there is an integer m such that $|Tf^m(v)| > 2|v|$. (Here we do not even need the number N.) Then there is m_2 such that $|Tf^{m_2}(v)| > 2|Tf^m(v)| > 4|v|$. Generally, for any positive integer k, there is an integer m_k such that $|Tf^{m_k}(v)| > 2^k|v|$. Thus the Tf-orbit of v is unbounded, proving Lemma 6.3. □

Let Λ be quasi-hyperbolic for f, and let N be the positive integer given by Lemma 6.3. We call $0 \neq u \in T_\Lambda M$ an *N-rightmax* if $|u| \geq |Tf^n(u)|$ for all $-N \leq n \leq 0$. Likewise, we call $0 \neq u \in T_\Lambda M$ an *N-leftmax* if $|u| \geq |Tf^n(u)|$ for all $0 \leq n \leq N$.

Thus, within any $2N+1$ consecutive iterates of any vector $0 \neq v \in T_\Lambda M$, there must be either an N-rightmax or an N-leftmax. (The maximum will be that one.)

Lemma 6.4. *Let Λ be quasi-hyperbolic for f. For any $0 \neq v \in T_\Lambda M$, if $Tf^i(v)$ is an N-leftmax and $Tf^j(v)$ is an N-rightmax on the same orbit of $v \neq 0$, then $i < j$.*

Proof. Otherwise, if $j \leq i$, then there would be $j \leq k \leq i$ such that $|Tf^k(v)|$ assumes the maximum on the whole interval $[j - N, i + N]$, contradicting Lemma 6.3. □

Let Λ be quasi-hyperbolic for f. Denote

$$\lambda = 2^{-1/N}, \ C = (\lambda^N \cdot \min\{|Tf^n(v)| \mid v \in T_\Lambda M, |v| = 1, 0 \leq n \leq N\})^{-2},$$

where N is given by Lemma 6.3.

Lemma 6.5. *Let Λ be quasi-hyperbolic for f, and let N, λ, and C be as just determined. If $u \in T_\Lambda M$ is an N-rightmax, then for any $i \geq 0$ and any $n \geq 0$, $|Tf^{i+n}(u)| \geq C^{-1}\lambda^{-n}|Tf^i(u)|$. Likewise, if $u \in T_\Lambda M$ is an N-leftmax, then for any $i \geq 0$ and any $n \geq 0$, $|Tf^{-i-n}(u)| \geq C^{-1}\lambda^{-n}|Tf^{-i}(u)|$.*

Proof. Let $u \in T_\Lambda M$ be an N-rightmax. By Lemmas 6.3 and 6.4, there is $0 \leq m_1 \leq N$ such that $Tf^{m_1}(u)$ is an N-rightmax and $|Tf^{m_1}(u)| \geq 2|u|$. Likewise, there is $0 \leq m_2 \leq N$ such that $Tf^{m_1+m_2}(u)$ is an N-rightmax and

$|Tf^{m_1+m_2}(u)| \geq 4|u|$, etc. A calculation similar to that done in the proof of Theorem 6.2 then shows that for any $n \geq 0$,

$$|Tf^n(u)| \geq C^{-1/2}\lambda^{-n}|u|.$$

Thus for any $i \geq 0$, with the help of the N-rightmax closest to $Tf^i(u)$, we get

$$|Tf^{i+n}(u)| \geq C^{-1}\lambda^{-n}|Tf^i(u)|$$

for any $n \geq 0$. The proof for N-leftmax is similar. This proves Lemma 6.5. □

Let Λ be quasi-hyperbolic for f, and let N be the positive integer given by Lemma 6.3. Denote

$$H^s = \{v \in T_\Lambda M - \{0\} \mid \{|Tf^n(v)|\}_{n=-\infty}^{\infty} \text{ has only } N\text{-leftmax}\},$$

$$H^u = \{v \in T_\Lambda M - \{0\} \mid \{|Tf^n(v)|\}_{n=-\infty}^{\infty} \text{ has only } N\text{-rightmax}\},$$

$$H^\vee = \{v \in T_\Lambda M - \{0\} \mid \{|Tf^n(v)|\}_{n=-\infty}^{\infty} \text{ has both } N\text{-leftmax and } N\text{-rightmax}\}.$$

This gives a Tf-invariant decomposition of $T_\Lambda M - \{0\}$. See Figure 6.1.

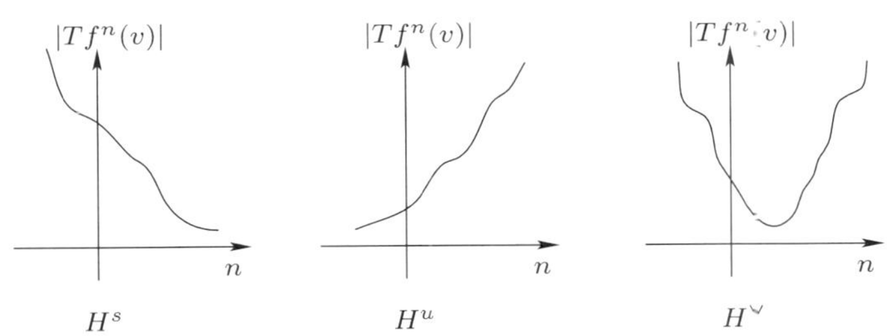

Figure 6.1. Norm of vectors under quasi-hyperbolicity.

That is, (the norms of the) vectors of $T_\Lambda M - \{0\}$ have three kinds of behavior under iterations. Roughly, it is either going down, or going up, or first going down and then going up. No one is first going up and then going down because no one has bounded orbit. The next corollary gives precise statements.

Corollary 6.6. *Let Λ be quasi-hyperbolic for f, and let N, λ, and C be determined as right before Lemma 6.5. Then:*

(1) *For any $v \in H^s$, $|Tf^n(v)| \leq C\lambda^n|v|$ for any $n \geq 0$.*

(2) *For any $v \in H^u$, $|Tf^{-n}(v)| \leq C\lambda^n|v|$ for any $n \geq 0$.*

(3) *For any $v \in H^\vee$, there is an integer i_0 such that for any $i \geq i_0$ and any $n \geq 0$, $|Tf^{i+n}(v)| \geq C^{-1}\lambda^{-n}|Tf^i(v)|$. Also, there is an integer j_0 such that for any $j \leq j_0$ and $n \geq 0$, $|Tf^{j-n}(v)| \geq C^{-1}\lambda^{-n}|Tf^j(v)|$.*

Proof. For any $v \in H^u$, there is an integer $i \geq 0$ such that $Tf^{-i}(v)$ is an N-rightmax. By Lemma 6.5, for any $n \geq 0$,

$$|Tf^n(v)| \geq C^{-1}\lambda^{-n}|v|.$$

Since H^u is Tf-invariant, this is equivalent to

$$|Tf^{-n}(v)| \leq C\lambda^n|v|.$$

This proves option (2). The other two options can be proved similarly. □

Theorem 6.7. *Let Λ be quasi-hyperbolic for f, and let λ and C be determined as right before Lemma 6.5. Then*

$$|Tf^n(v)| \leq C\lambda^n|v|, \ \forall v \in B^s(x), x \in \Lambda, n \geq 0,$$
$$|Tf^{-n}(v)| \leq C\lambda^n|v|, \ \forall v \in B^u(x), x \in \Lambda, n \geq 0.$$

Proof. We only need to prove that $B^s - \{0\} \subset H^s$. This is obvious because, for any $0 \neq v \in B^s$, there can be no N-rightmax by Lemma 6.5. The proof for B^u is similar. □

We remark that $H^s \subset B^s$ by option (1) of Corollary 6.6. Hence $H^s \subset B^s - \{0\}$. Likewise for H^u. Thus

$$B^s - \{0\} = H^s, \ B^u - \{0\} = H^u.$$

This gives a characterization for hyperbolicity:

Theorem 6.8. *A compact invariant set Λ of f is hyperbolic if and only if $B^s(x) \oplus B^u(x) = T_xM$ for any $x \in \Lambda$.*

Proof. We only prove the "if" part. Assume $B^s(x) \oplus B^u(x) = T_xM$ for any $x \in \Lambda$. Since Λ is compact, Λ is quasi-hyperbolic for f. Then the conclusion follows immediately from Theorem 6.7. □

Let Λ be quasi-hyperbolic for f. For $i = 0, 1, \dots, \dim(M)$, let

$$\Delta^i = \Delta^i(\Lambda) = \{x \in \Lambda \mid B^s(x) \oplus B^u(x) = T_xM, \ \dim B^s(x) = i\}.$$

Clearly, Δ^i is f-invariant. Since Λ is quasi-hyperbolic for f, by Theorem 6.7, Δ^i is hyperbolic for f of index i. Also, the same proof as that of Theorem 4.3 (the uniform (C, λ) rates pass to the closure) shows Δ^i is compact. Clearly, $\Delta^i \cap \Delta^j = \emptyset$ if $i \neq j$.

Let

$$\Delta = \bigcup_{i=0}^{\dim M} \Delta^i.$$

Then Δ is f-hyperbolic. We prove in Theorems 6.9 through 6.11 that Δ contains the chain recurrent set of $f|_\Lambda$.

Theorem 6.9. *Let Λ be quasi-hyperbolic for f. For any $x \in \Lambda$, $\omega(x)$ is hyperbolic of index* $\dim(B^s(x))$ *with contraction rates (C, λ) determined as right before Lemma 6.5. Likewise, $\alpha(x)$ is hyperbolic of index* $\dim(M) - \dim(B^u(x))$ *with the same rates (C, λ).*

Proof. Let $x \in \Lambda$ and $y \in \omega(x)$. Take a subspace $F(x)$ of $E(x)$ such that

$$F(x) \oplus B^s(x) = T_x M.$$

Since $B^s(x) - \{0\} = H^s(x)$ and since $H^s(x) \cup H^u(x) \cup H^\vee(x)$ is a decomposition of $T_x M - \{0\}$, for any $v \in F(x)$ with $|v| = 1$, there is a positive integer $i_0 = i_0(v)$ such that $Tf^{i_0}(v)$ is an N-rightmax. By Lemma 6.3, there is $i_1 = i_1(v)$ with $i_0 \le i_1 \le i_0 + N$ such that $Tf^{i_1}(v)$ is a "strict" N-rightmax in the sense that $|Tf^{i_1}(v)| > |Tf^n(v)|$ for all $i_1 - N \le n \le i_1 - 1$. Then there is a neighborhood U of v in the unit sphere of $F(x)$ such that $Tf^{i_1}(w)$ is an N-rightmax for all $w \in U$. By compactness there is a positive integer m_0 such that for all $v \in F(x)$ with $|v| = 1$, there is $m = m(v)$ with $0 \le m \le m_0$ such that $Tf^m(v)$ is an N-rightmax. By Lemma 6.5,

$$|Tf^{i+n}(v)| \ge C^{-1}\lambda^{-n}|Tf^i(v)|$$

for any $v \in F(x)$, $i \ge m_0$, and $n \ge 0$. Also, by Theorem 6.7,

$$|Tf^{i+n}(v)| \le C\lambda^n |Tf^i(v)|$$

for any $v \in B^s(x)$, $i \in \mathbb{Z}$, and $n \ge 0$.

Let $f^{n_k}(x), n_k \to \infty$, be a sequence that tends to y such that $Tf^{n_k}(B^s(x))$ and $Tf^{n_k}(F(x))$ tend to two subspaces $G^s(y)$ and $G^u(y)$ of $T_y M$, respectively. For any $w \in G^s(y)$, there is a sequence $w_k \in Tf^{n_k}(B^s(x))$ with $w_k \to w$. Since

$$|Tf^n(w_k)| \le C\lambda^n |w_k|$$

for any $k \ge 1$ and $n \ge 0$, fixing n and letting $k \to \infty$ yield

$$|Tf^n(w)| \le C\lambda^n |w|, \ \forall w \in G^s(y), \ n \ge 0.$$

Thus

$$G^s(y) \subset B^s(y).$$

Likewise, for any $w \in G^u(y)$, there is a sequence $w_k \in Tf^{n_k}(F(x))$ with $w_k \to w$. Fix $j \ge 0$. Consider those n_k in the sequence with $n_k - j > m_0$. Then

$$|w_k| \ge C^{-1}\lambda^{-j}|Tf^{-j}(w_k)|.$$

Letting $k \to \infty$ yields

$$|w| \ge C^{-1}\lambda^{-j}|Tf^{-j}(w)|, \ \forall w \in G^u(y).$$

See Figure 6.2. Since $j \ge 0$ can be arbitrary, this gives

$$G^u(y) \subset B^u(y).$$

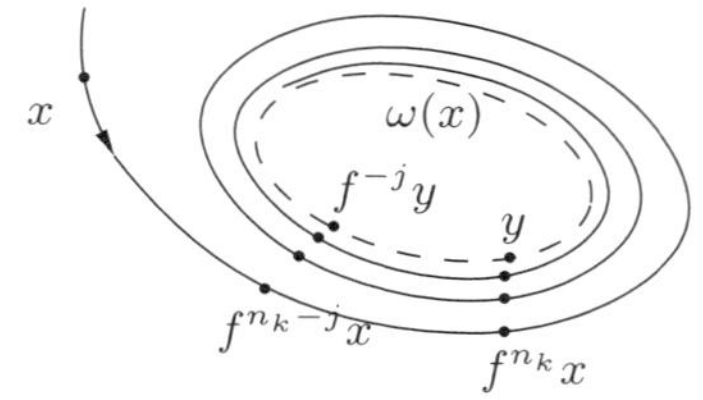

Figure 6.2. The proof of Theorem 6.9. Vectors of $F(x)$ will start to expand from any moment that is after m_0; hence it will be expanding for the stride from $n_k - j$ to n_k if $n_k - j > m_0$. Passing to the limit, vectors of $G^u(y)$ will be expanding for the stride from $-j$ to 0.

Counting dimensions we get

$$B^s(y) \oplus B^u(y) = T_y M.$$

This proves that $\omega(x)$ is f-hyperbolic of index $\dim B^s(x)$ with contraction rates (C, λ). The proof for $\alpha(x)$ is similar. This proves Theorem 6.9. □

Theorem 6.10. *If Λ is quasi-hyperbolic for f, then f restricted to $L(f|_\Lambda)$ is hyperbolic.*

Proof. Because the constants (C, λ) determined as right before Lemma 6.5 are independent of $\omega(x)$, $x \in \Lambda$, the inequalities carry over to the closure. Then Theorem 6.10 follows from Theorem 6.9. □

Theorem 6.10 can be extended to the following.

Theorem 6.11. *If Λ is quasi-hyperbolic for f, then f restricted to* $\mathrm{CR}(f|_\Lambda)$ *is hyperbolic.*

Proof. By Theorem 6.10, $L(f|_\Lambda)$ is contained in Δ. If we prove there is no cycle between the $\{\Delta^i\}$ with respect to $f|_\Lambda$, then, by Theorem 5.9, Δ will contain $\mathrm{CR}(f|_\Lambda)$, proving Theorem 6.11.

Suppose there is a cycle, say

$$x_i \in W^u(\Delta^{k_i}, f|_\Lambda) \cap W^s(\Delta^{k_{i+1}}, f|_\Lambda) - \Delta, \ i = 1, \dots, m,$$

where $\Delta^{k_{m+1}} = \Delta^{k_1}$. By Theorem 6.9,

$$\dim B^s(x_i) = k_{i+1}, \ \dim B^u(x_i) = \dim M - k_i$$

for all $i = 1, \dots, m$. Hence

$$\dim B^s(x_i) + \dim B^u(x_{i+1}) = \dim M$$

for each i. On the other hand, since Λ is quasi-hyperbolic,

$$B^s(x_i) \cap B^u(x_i) = \{0\}$$

for each i. In particular,

$$\dim B^s(x_i) + \dim B^u(x_i) \leq \dim M$$

for each i. Counting dimensions then shows

$$\dim B^s(x_i) + \dim B^u(x_i) = \dim M$$

for each i. Hence

$$B^s(x_i) \oplus B^u(x_i) = T_{x_i}M$$

for each i. Thus $x_i \in \Delta$, a contradiction. This proves there is no cycle between the $\{\Delta^i\}$ with respect to $f|_\Lambda$. $\square$

Recall that the *dual bundle* $T^*_\Lambda M$ of $T_\Lambda M$ has fiber $T^*_x M$, the cotangent space of M at $x \in \Lambda$, which consists of linear functionals of $T_x M$. The *dual map* $T^*f : T^*_\Lambda M \to T^*_\Lambda M$ of Tf is defined by

$$(T^*f(\alpha))(v) = \alpha(Tf(v))$$

for any $v \in T_x M$, $x \in \Lambda$, and $\alpha \in T^*_{fx}M$. Thus T^*f is a bundle isomorphism of $T^*_\Lambda M$ over f^{-1}.

The notion of quasi-hyperbolicity can be defined for T^*f on $T^*_\Lambda M$ in the same way as we did for Tf on $T_\Lambda M$. That is, T^*f on $T^*_\Lambda M$ is called *quasi-hyperbolic* if

$$B^s(x, T^*f) \cap B^u(x, T^*f) = \{0\}$$

for every $x \in \Lambda$, where

$$B^s(x, T^*f) = \{\alpha \in T^*_x M \mid \{|(T^*f)^n(\alpha)|\}_{n=0}^{\infty} \text{ is bounded}\},$$

$$B^u(x, T^*f) = \{\alpha \in T^*_x M \mid \{|(T^*f)^{-n}(\alpha)|\}_{n=0}^{\infty} \text{ is bounded}\}.$$

We say that T^*f on $T^*_\Lambda M$ is quasi-hyperbolic instead of saying that "Λ is quasi-hyperbolic". This is because by "Λ is quasi-hyperbolic" we already mean that Tf on $T_\Lambda M$ is quasi-hyperbolic. To avoid confusion, for quasi-hyperbolicity of T^*f we will not refer to the *set* Λ, but to "T^*f on $T^*_\Lambda M$" explicitly.

All results of Section 6.2 hold for the dual setting $(T^*f, T^*_\Lambda M, \Lambda, f^{-1})$. For instance, Theorem 6.11 in this setting will read "*If T^*f on $T^*_\Lambda M$ is quasi-hyperbolic, then f^{-1} restricted to* $\mathrm{CR}(f^{-1}|_\Lambda)$, *the chain recurrent set of the restricted map, is hyperbolic.*" It is straightforward to check all the proofs.

6.3. Linear transversality

For $x \in \Lambda$, denote

$$D^s(x) = \{v \in T_xM \mid |Tf^n(v)| \to 0,\ n \to \infty\},$$

$$D^u(x) = \{v \in T_xM \mid |Tf^{-n}(v)| \to 0,\ n \to \infty\}.$$

These are linear subspaces of T_xM, invariant under Tf, as family. We say Tf on $T_\Lambda M$ satisfies the *linear transversality condition* if

$$D^s(x) + D^u(x) = T_xM$$

for all $x \in \Lambda$.

The linear transversality condition is introduced by Robbin (1971). Mañé (1977b) links it beautifully to quasi-hyperbolicity of the dual isomorphism T^*f.

For any linear subspace $F(x)$ of T_xM, define the *annihilator* of $F(x)$ to be

$$(F(x))^0 = \{\alpha \in T_x^*M \mid \alpha(v) = 0 \text{ for all } v \in F(x)\}.$$

This is a linear subspace of T_x^*M. Note that

$$\dim(F(x))^0 = \dim M - \dim F(x).$$

It is straightforward to check that, for any two linear subspaces $F(x)$ and $G(x)$ of T_xM,

$$F(x) \cap G(x) = \{0\} \Longleftrightarrow (F(x))^0 + (G(x))^0 = T_x^*M$$

and

$$F(x) + G(x) = T_xM \Longleftrightarrow (F(x))^0 \cap (G(x))^0 = \{0\}.$$

We define the hyperbolicity of T^*f on T_Λ^*M in the obvious way by saying that T^*f on T_Λ^*M is *hyperbolic* if, for each $x \in \Lambda$, the cotangent space T_x^*M splits into a direct sum

$$T_x^*M = \Gamma^s(x) \oplus \Gamma^u(x),$$

invariant (as family) in the sense that

$$T^*f(\Gamma^s(x)) = \Gamma^s(f^{-1}(x)),\ T^*f(\Gamma^u(x)) = \Gamma^u(f^{-1}(x))$$

such that, for some constants $C \geq 1$ and $0 < \lambda < 1$, the following uniform estimates hold:

$$|(T^*f)^n(\alpha)| \leq C\lambda^n|\alpha|,\ \forall x \in \Lambda,\ \alpha \in \Gamma^s(x),\ n \geq 0,$$

$$|(T^*f)^{-n}(\alpha)| \leq C\lambda^n|\alpha|,\ \forall x \in \Lambda,\ \alpha \in \Gamma^u(x),\ n \geq 0.$$

Theorem 6.12. *Let $\Lambda \subset M$ be a compact invariant set of f. Then Λ is hyperbolic for f if and only if T^*f on T_Λ^*M is hyperbolic.*

Proof. Let $T_\Lambda M = E^s \oplus E^u$ be the hyperbolic splitting of f with constants $0 < \lambda < 1$ and $C \geq 1$. Then the splitting

$$(E^s)^0 \oplus (E^u)^0 = T_\Lambda^* M$$

is T^*f-invariant. Since the angles between $E^s(x)$ and $E^u(x)$ have a positive lower bound, there is $K > 0$ such that

$$|v^s| \leq K|v|$$

for all $v \in T_\Lambda M$, where $v = v^s + v^u$, $v^s \in E^s$, $v^u \in E^u$. Then for any $\alpha \in (E^u(x))^0$ and $v \in T_{f^{-n}x}M$,

$$|(T^*f)^n(\alpha)(v)| = |\alpha(Tf^n(v))| = |\alpha(Tf^n(v^s))| \leq C\lambda^n|\alpha||v^s| \leq KC\lambda^n|\alpha||v|.$$

Hence

$$|(T^*f)^n(\alpha) \leq KC\lambda^n|\alpha|.$$

This proves that $(E^u)^0$ is contracting under T^*f. Likewise $(E^s)^0$ is expanding under T^*f. This proves Theorem 6.12. $\square$

Theorem 6.13. *Let $\Lambda \subset M$ be a compact invariant set of f. Then Tf on $T_\Lambda M$ satisfies the linear transversality condition if and only if T^*f on T_Λ^*M is quasi-hyperbolic.*

Proof. First we give the "only if" part. Let Tf on $T_\Lambda M$ satisfy the linear transversality condition. Let $x \in \Lambda$ and $\alpha \in T_x^*M$ satisfy

$$|(T^*f)^n(\alpha)| \leq K$$

for all integers n. Take any $v \in T_xM$. By the linear transversality condition, there are $v^s \in D^s(x)$ and $v^u \in D^u(x)$ such that

$$v = v^s + v^u.$$

Then

$$|\alpha(v^s)| = |\alpha(Tf^{-n}Tf^n(v^s))| = |((T^*f)^{(-n)}\alpha)(Tf^n(v^s))| \leq K|Tf^n(v^s)| \to 0.$$

Hence $\alpha(v^s) = 0$. Likewise $\alpha(v^u) = 0$. Thus $\alpha(v) = 0$ for all $v \in T_xM$. Hence $\alpha = 0$. This proves that T^*f on T_Λ^*M is quasi-hyperbolic.

Now we give the "if" part. Let T^*f on T_Λ^*M be quasi-hyperbolic. We prove

$$D^s(x) + D^u(x) = T_xM$$

for all $x \in \Lambda$. This is equivalent to proving that

$$(D^s(x))^0 \cap (D^u(x))^0 = \{0\}$$

for all $x \in \Lambda$. Since T^*f on T_Λ^*M is quasi-hyperbolic, or equivalently,

$$B^s(x, T^*f) \cap B^u(x, T^*f) = \{0\},$$

it suffices to prove

$$(D^s(x))^0 \subset B^u(x, T^*f),\ (D^u(x))^0 \subset B^s(x, T^*f)$$

for all $x \in \Lambda$. We prove the first $\subset$. The second can be proved similarly.

Let $x \in \Lambda$. Since $L(f^{-1}|_\Lambda) = L(f|_\Lambda)$, by Theorem 6.10, T^*f restricted to $L(f|_\Lambda)$ is hyperbolic. By Theorem 6.12, Tf restricted to $L(f|_\Lambda)$ is hyperbolic. Then $\omega(x)$ is hyperbolic for f. Take a subspace $F(x)$ of T_xM such that

$$F(x) \oplus D^s(x) = T_xM.$$

Then if $f^{n_k}(x) \to y \in \omega(x)$ such that $Tf^{n_k}(F(x))$ tends to a subspace $F(y)$ of T_yM and $Tf^{n_k}(D^s(x))$ tends to a subspace $D(y)$ of T_yM, then

$$F(y) = E^u(y),\ D(y) = E^s(y).$$

This implies there are $N \geq 1$ and $m \geq 1$ large such that for every $n \geq m$,

$$|Tf^{-n}|_{Tf^{N+n}(F(x))}| \leq 1.$$

Hence there is $L > 0$ such that

$$|Tf^{-n}|_{Tf^n(F(x))}| \leq L$$

for all $n \geq 0$. Also, there is $K > 0$ such that

$$|\pi_n| \leq K$$

for all $n \geq 0$, where

$$\pi_n : T_{f^n x}M \to Tf^n(F(x))$$

is the projection along $Tf^n(D^s(x))$.

Now let $\alpha \in (D^s(x))^0$. For any $n \geq 0$ and $v \in T_{f^n x}M$,

$$\begin{aligned} |(T^*f^{-n}(\alpha))(v)| &= |\alpha(Tf^{-n}(v))| = |\alpha(Tf^{-n}(\pi_n v))| \\ &\leq LK|\alpha||v|. \end{aligned}$$

Hence

$$|T^*f^{-n}(\alpha)| \leq LK|\alpha|.$$

This proves $\alpha \in B^u(x, T^*f)$ and completes the proof of Theorem 6.13. □

6.4. Applications

We give some equivalent conditions for compact hyperbolic sets.

Theorem 6.14. *Let $\Lambda \subset M$ be a compact invariant set of f. The following conditions are equivalent:*

(1) *Λ is hyperbolic for f.*

(2) *$B^s(x) \oplus B^u(x) = T_xM$ for all $x \in \Lambda$.*

(3) *$D^s(x) \oplus D^u(x) = T_xM$ for all $x \in \Lambda$.*

Proof. That (1) $\Leftrightarrow$ (2) is just Theorem 6.8. Obviously (1) $\Rightarrow$ (3). It remains to prove (3) $\Rightarrow$ (1).

Assume condition (3). By Theorem 6.13, T^*f on $T^*_\Lambda M$ is quasi-hyperbolic; that is,

$$B^s(x, T^*f) \cap B^u(x, T^*f) = \emptyset$$

for all $x \in \Lambda$. Since condition (3) is equivalent to

$$(D^s(x))^0 \oplus (D^u(x))^0 = T^*_x M$$

and since in the proof of Theorem 6.13 we proved

$$(D^s(x))^0 \subset B^u(x, T^*f), \ (D^u(x))^0 \subset B^s(x, T^*f),$$

it follows that

$$B^s(x, T^*f) \oplus B^u(x, T^*f) = T^*_x M$$

for all $x \in \Lambda$. By Theorem 6.8, T^*f on $T^*_\Lambda M$ is hyperbolic. By Theorem 6.12, Λ is hyperbolic for f. This proves Theorem 6.14. □

Denote by $\mathcal{C}(M)$ the set of nonempty compact subsets of M. For $X, Y \in \mathcal{C}(M)$, define

$$d_H(X, Y) = \inf\{\epsilon \mid X \subset B(Y, \epsilon), \ Y \subset B(X, \epsilon)\},$$

where, as usual,

$$B(X, \epsilon) = \{x \in M \mid d(x, X) \leq \epsilon\}.$$

Then d_H is a metric on $\mathcal{C}(M)$, called the *Hausdorff metric*. It is standard that, with respect to the Hausdorff metric, $\mathcal{C}(M)$ is compact. For a reference the reader is referred to Munkres (2000).

We insert a lemma saying that chain recurrent points are chain recurrent in their own right:

Lemma 6.15. $\mathrm{CR}(f|_{\mathrm{CR}(f)}) = \mathrm{CR}(f)$.

Proof. We prove $\mathrm{CR}(f) \subset \mathrm{CR}(f|_{\mathrm{CR}(f)})$. The other direction is obvious.

Let $x \in \mathrm{CR}(f)$ be given. For every $n \geq 1$, there is a periodic $\frac{1}{n}$-chains

$$C_n = \{x_0^n, x_1^n, \ldots, x_{j_n}^n\}$$

such that $x_0^n \to x$. Taking a subsequence if necessary, we may assume the C_n converge to a compact set K in the Hausdorff metric. Clearly

$$K \subset \mathrm{CR}(f),$$

and $x \in K$. It suffices to prove that for any $\epsilon > 0$, there is an ϵ-chain $\{y_i\} \subset K$ going from and back to the ϵ-neighborhood of x. Let $\delta > 0$ be small such that $d(p, q) < \delta$ implies

$$d(f(p), f(q)) < \epsilon/3$$

for any $p, q \in M$. We may assume $\delta \leq \epsilon/3$. Take N large such that

$$d(x_{i+1}^n, f(x_i^n)) \leq \epsilon/3$$

for all $n \geq N$ and $0 \leq i \leq j_n$. Fix an integer $n \geq N$ such that

$$d(x_0^n, x) \leq \delta$$

and

$$d_H(C_n,\ K) \leq \delta,$$

where d_H denotes the Hausdorff metric. Then for each $i = 0, \ldots, j_n$, there is $y_i \in K$ such that

$$d(y_i, x_i^n) \leq \delta.$$

We verify that $\{y_0, y_1, \ldots, y_{j_n}\}$ satisfies the requirement. In fact, y_0 and y_{j_n} are in $B(x_0^n, \delta) \subset B(x, \epsilon)$. Moreover, for each $0 \leq i \leq j_n - 1$,

$$\begin{aligned} d(y_{i+1}, fy_i) &\leq d(y_{i+1}, x_{i+1}^n) + d(x_{i+1}^n, f(x_i^n)) + d(f(x_i^n), fy_i) \\ &\leq \delta + \epsilon/3 + \epsilon/3 \leq \epsilon. \end{aligned}$$

This proves Lemma 6.15. □

Recall from Theorem 5.11 that Axiom A plus no-cycle condition is equivalent to CR(f) being hyperbolic. Here are more equivalent conditions.

Theorem 6.16. *The following conditions are equivalent:*

(1) CR(f) *is hyperbolic.*

(2) CR(f) *is quasi-hyperbolic.*

(3) CR(f) *satisfies the linear transversality condition.*

Note that, in our use of terminologies, condition (2) means Tf on CR(f) is quasi-hyperbolic, and condition (3) means Tf on CR(f) satisfies the linear transversality condition.

Proof. That (1) ⇒ (2) is obvious. We prove (2) ⇒ (1). Assume f is quasi-hyperbolic on CR(f). By Theorem 6.11, CR$(f|_{\mathrm{CR}(f)})$ is hyperbolic for f. By Lemma 6.15, CR(f) is hyperbolic for f, proving (2) ⇒ (1).

Also, (1) ⇒ (3) is obvious. It remains to prove (3) ⇒ (1). Let f satisfy the linear transversality condition on CR(f). By Theorem 6.13, T^*f on $T^*_{\mathrm{CR}(f)}M$ is quasi-hyperbolic. By Theorem 6.11, T^*f on $T^*_{\mathrm{CR}(f^{-1}|_{\mathrm{CR}(f^{-1})})}M = T^*_{\mathrm{CR}(f|_{\mathrm{CR}(f)})}M$ is hyperbolic. By Theorem 6.12, CR$(f|_{\mathrm{CR}(f)})$ is hyperbolic for f. By Lemma 6.15, CR(f) is hyperbolic for f, proving (3) ⇒ (1). This proves Theorem 6.16. □

We consider the special case when $\Lambda = M$. A diffeomorphism $f : M \to M$ is called *quasi-Anosov* if Tf on TM is quasi-hyperbolic, and it is said to *satisfy the linear transversality condition* if Tf on TM satisfies the linear transversality condition.

These two conditions do not imply one another. The north-south poles map of S^2 satisfies the linear transversality condition but is not quasi-Anosov. Franks and Robinson (1976) give a diffeomorphism that is quasi-Anosov but not Anosov. The example also shows that being quasi-Anosov does not imply the linear transversality condition because a diffeomorphism that is simultaneously quasi-Anosov and satisfying the linear transversality condition must be Anosov.

Theorem 6.17. *f is quasi-Anosov if and only if f satisfies Axiom A and $T_x(W^s(x)) \cap T_x(W^u(x)) = \{0\}$ for all $x \in M$.*

Proof. Let f satisfy Axiom A, and let

$$T_x(W^s(x)) \cap T_x(W^u(x)) = \{0\}$$

for all $x \in M$. By Theorem 5.5,

$$B^s(x) = T_x(W^s(x)),\ B^u(x) = T_x(W^u(x))$$

for all $x \in M$. Now

$$T_x(W^s(x)) \cap T_x(W^u(x)) = \{0\}$$

for all $x \in M$; hence f is quasi-Anosov.

Conversely, let f be quasi-Anosov. By Theorem 6.11, $\mathrm{CR}(f)$ is hyperbolic. By Theorem 5.11, f satisfies Axiom A (and the no-cycle condition). By Theorem 5.5,

$$B^s(x) = T_x(W^s(x)),\ B^u(x) = T_x(W^u(x))$$

for all $x \in M$. Now f is quasi-Anosov; hence

$$T_x(W^s(x)) \cap T_x(W^u(x)) = \{0\}$$

for all $x \in M$. This proves Theorem 6.17. □

Theorem 6.18. *f satisfies the linear transversality condition if and only if f satisfies Axiom A and the strong transversality condition.*

Proof. Let f satisfy Axiom A and the strong transversality condition. By Theorem 5.5,

$$D^s(x) = T_x(W^s(x)),\ D^u(x) = T_x(W^u(x))$$

for all $x \in M$. Thus f satisfies the linear transversality condition.

Conversely, let f satisfy the linear transversality condition. By Theorem 6.13, $T^*f : T^*M \to T^*M$ is quasi-Anosov. By Theorems 6.11 and 6.12, $\mathrm{CR}(f)$ is hyperbolic for f. By Theorem 5.11, f satisfies Axiom A. By Theorem 5.5,

$$D^s(x) = T_x(W^s(x)),\ D^u(x) = T_x(W^u(x))$$

for all $x \in M$. Now f satisfies the linear transversality condition; hence f satisfies the strong transversality condition. This proves Theorem 6.18. □

We put these results into the following diagram. See Figure 6.3. The two arrows at the bottom are left as exercises.

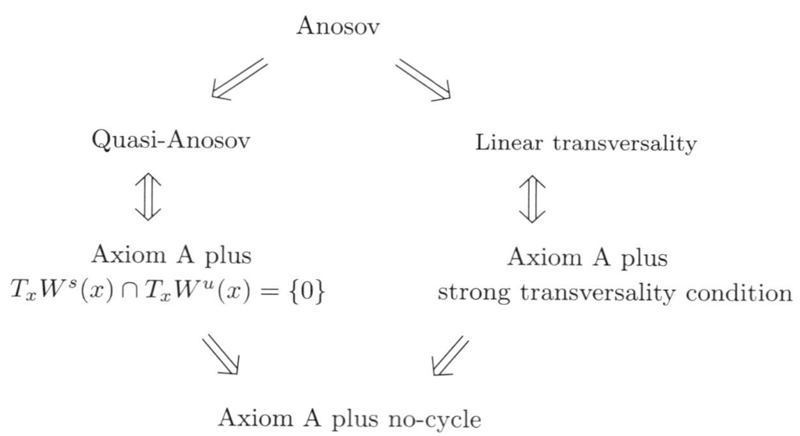

Figure 6.3. A summary diagram.

6.5. A glimpse of the stability conjectures

The main result of this text, the Ω-stability theorem of Smale (Theorem 5.8), is part of the problem of characterizing structural stability, known as the stability conjectures stated next, proposed in a joint work of Palis and Smale (1970), which remained a central problem in differentiable dynamical systems for several decades. In this section we give a glimpse of these conjectures. It is like a bird's eye view without details. We state the results for diffeomorphisms. Then we briefly mention the story for flows.

The stability conjecture. *f is C^r structurally stable if and only if f satisfies Axiom A and the strong transversality condition.*

The Ω-stability conjecture. *f is C^r Ω-stable if and only if f satisfies Axiom A and the no-cycle condition.*

The Ω-stability theorem of Smale (Theorem 5.8) corresponds to the "if" part of the second conjecture. The "if" part of the first conjecture is the structural stability theorem (Theorem 5.16), which demands control of not only the nonwandering orbits, but also the wandering orbits. After a number of previous works, notably Palis (1968), Palis-Smale (1970), Robbin (1971), and de Melo (1973), it was eventually solved by Robinson (1976), as stated at the end of Chapter 5.

The "only if" part of the two conjectures first went through some reductions. Palis (1970) proved that *if f is C^r Ω-stable and satisfies Axiom A, then f satisfies the no-cycle condition.* Robinson (1973) proved that *if f is C^r structurally stable and satisfies Axiom A, then f satisfies the strong transversality condition.* Thus the two "only if" parts both reduce to proving Axiom A:

The stability conjecture (reduced version). *If f is C^r structurally stable, then f satisfies Axiom A.*

The Ω-stability conjecture (reduced version). *If f is C^r Ω-stable, then f satisfies Axiom A.*

There is an abuse of language here; i.e., part of a conjecture took the name of the whole conjecture. Nevertheless terminologies have been used this way because of the two reductions. Note that, stated this way, the Ω-stability conjecture implies the stability conjecture.

The two conjectures remained unsolved for a long time and received great attention in differentiable dynamical systems. Note that C^r stability implies C^{r+1} stability; hence the C^1 stability is the strongest. In the early 1980s Liao (1980a), Mañé (1982), and Sannami (1983) independently solved the C^1 stability conjecture for the 2-dimensional case. Several years later, Mañé (1988) solved the C^1 stability conjecture for general dimensions:

Theorem. *If f is C^1 structurally stable, then f satisfies Axiom A.*

Based on Mañé's results, Palis (1988) proved the C^1 Ω-stability conjecture:

Theorem. *If f is C^1 Ω-stable, then f satisfies Axiom A.*

Combined with the reduction mentioned earlier, C^1 Ω-stability actually implies Axiom A plus no-cycle. Recall that in Section 5.2 we mentioned without proof a striking fact that Ω-stability is equivalent to ϵ-Ω-stability, for the C^1 topology. This is a direct consequence of Theorem 5.8 and this result of Palis. Indeed, by (the proof of) Theorem 5.8, Axiom A plus no-cycle implies C^1 ϵ-Ω-stability, which implies C^1 Ω-stability. Now by Palis, C^1 Ω-stability implies Axiom A plus no-cycle. Thus C^1 Ω-stability is equivalent to C^1 ϵ-Ω-stability. Likewise for structural stability.

During the long march towards the solutions of the stability conjectures, Pliss, Liao, and Mañé noticed since the 1970s an important class of systems, the *star systems* (named by Liao). A diffeomorphism f is called a *star diffeomorphism* if there is a C^1 neighborhood $\mathcal{U}$ of f such that every periodic orbit of every $g \in \mathcal{U}$ is hyperbolic. Mañé (1982) proposed the following conjecture:

The star conjecture. *Every star diffeomorphism satisfies Axiom A and the no-cycle condition.*

Even earlier, Liao (1979b) proposed this conjecture for flows of lower dimensions. Also see Liao (1981, 1986).

The star condition is rather weak. It mentions periodic orbits only, and the hyperbolicity mentioned for periodic orbits is orbitwise, but not uniform. Indeed, the star condition follows from the C^1 Ω-stability easily. Let us explain. Assume f is C^1 Ω-stable. Then f has a C^1 neighborhood $\mathcal{U}$ such that all $g \in \mathcal{U}$ are Ω-conjugate. Since $\mathcal{U}$ contains some Kupka-Smale system that has at most countably many periodic points (this is the celebrated Kupka-Smale theorem; see for instance Robinson (1995)), every $g \in \mathcal{U}$ has at most countably many periodic points. On the other hand, if f is not a star system, some arbitrarily small C^1 perturbation will create a nonhyperbolic periodic point, say a nonhyperbolic fixed point of an eigenvalue one. By a lemma of Franks (1971), a further C^1 perturbation will create an arc of fixed points on the manifold, which is a contradiction.

Thus the star conjecture of Mañé and Liao serves as the third and the strongest version of the stability conjecture. The 2-dimensional case of the star conjecture was proved by Liao (1980a) and Mañé (1982). The general case was proved by Aoki (1992) and Hayashi (1992):

Theorem. *Every star diffeomorphism satisfies Axiom A and the no-cycle condition.*

Corollary. *The following three conditions are equivalent:*

(1) *f satisfies Axiom A plus the no-cycle condition.*

(2) *f is C^1 Ω-stable.*

(3) *f is a star diffeomorphism.*

That (1) $\Rightarrow$ (2) is the Ω-stability theorem of Smale (1970). That (2) $\Rightarrow$ (3) is by the Kupka-Smale theorem and a lemma of Franks (1971) as explained above. That (3) $\Rightarrow$ (1) is by Aoki (1992) and Hayashi (1992).

Thus, to establish the global Axiom A, now one only needs to verify the star condition, that is, to argue that if a nonhyperbolic periodic point, say a nonhyperbolic fixed point of an eigenvalue one, is created by perturbation, then one can create further by perturbation something strange that leads to a contradiction.

For simplicity we have not talked about flows (or vector fields), except in an informal survey at the very beginning of Chapter 1. In fact all the problems of the stability conjectures have corresponding versions for flows stated next.

Let X be a vector field on M. Pugh and Shub (1970) proved the Ω-stability theorem for flows:

Theorem. *If X satisfies Axiom A plus the no-cycle condition, then X is C^r Ω-stable for any $r \geq 1$.*

Robinson (1974, 1975) proved the structural stability theorem for flows:

Theorem. *If X satisfies Axiom A plus the strong transversality condition, then X is C^r structurally stable for any $r \geq 1$.*

Here the terminologies of structural stability, Axiom A, etc., as well as the statements, all look the same as those for diffeomorphisms. However, between flows and diffeomorphisms, there is an important difference behind the meaning of Axiom A or, more precisely, the meaning of hyperbolicity. Let us look at the notion of hyperbolic periodic orbit.

For diffeomorphisms, the hyperbolicity of a periodic orbit means that the tangent space of every point of the orbit splits into a direct sum of contracting and expanding subspaces. This is the same as the hyperbolicity of a fixed point. In particular, it is perfect for an Axiom A diffeomorphism to have hyperbolic periodic orbits accumulating on a hyperbolic fixed point, as shown by the Smale horseshoe.

However, for flows, the meaning of hyperbolicity for a periodic orbit is substantially different from the meaning of hyperbolicity for a singularity (rest point). For a singularity, hyperbolicity still means the splitting of the whole tangent space into contracting and expanding directions. See Figure 6.4. However, for a hyperbolic periodic orbit the flow direction along the orbit is neither contracting nor expanding, but neutral. It is the codimension 1 normal space of the flow direction that splits into contracting and expanding directions. It is actually a *normal hyperbolicity*. See Figure 6.4. Thus hyperbolic singularities and hyperbolic periodic orbits, though both named "hyperbolic", are two different notions. Indeed, for flows, the definition of Axiom A has an extra requirement that periodic orbits do not accumulate on singularities. See Smale (1967). In other words, contrary to diffeomorphisms, for flows accumulation of periodic orbits on singularities violates Axiom A.

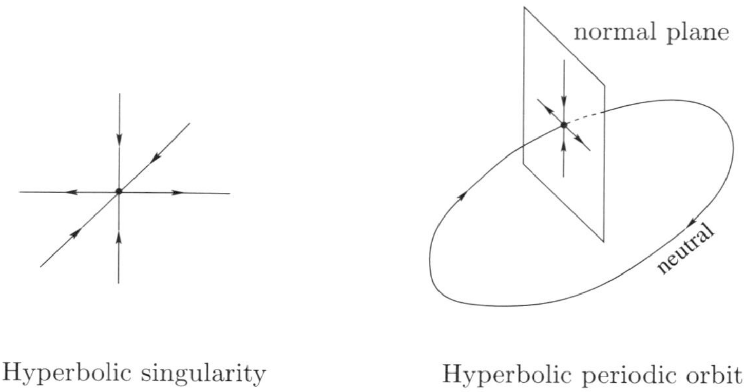

Figure 6.4. Hyperbolic singularity and periodic orbit for flows.

Now to prove the stability conjecture we are assuming structural stability to prove Axiom A. Thus, for flows, one has to prove for structurally stable systems that periodic orbits do not accumulate on singularities, which is a hard problem. For diffeomorphisms this is simply not an issue.

There are other important differences between flows and diffeomorphisms, even between nonsingular flows and diffeomorphisms. Some basic methods used for diffeomorphisms do not carry over to flows. This yields a special concern about the characteristics of flows. In this spirit, with a goal of structural stability, a fundamental study was carried out by Liao (1996).

In 1997, Hayashi proved the C^1 connecting lemma and the C^1 stability conjecture and Ω-stability conjecture for flows:

Theorem. *If X is C^1 structurally stable or C^1 Ω-stable, then X satisfies Axiom A.*

Before that, assuming a C^1 connecting lemma, Wen (1996) was able to prove the C^1 stability conjecture for flows. Even earlier, Hu (1994) proved the C^1 stability conjecture for flows in dimension 3.

The star conjecture for flows is also settled. Liao (1981) proved the 3-dimensional case. Gan-Wen (2006) proved the n-dimensional case:

Theorem. *Every separated star flow satisfies Axiom A and the no-cycle condition.*

Here a star flow X is called *separated* if there are a C^1 neighborhood $\mathcal{U}$ of X and a neighborhood U of $\text{Sing}(X)$ in M such that every periodic orbit of every $Y \in \mathcal{U}$ is contained in $M - U$. This separation assumption is proposed by Liao (1981), which turns out to be sharp (see the next corollary). Mañé (1982) points out that without separation the conjecture will be false: the geometrical Lorenz attractor is a nonseparated star flow that does not satisfy Axiom A. More counterexamples can be found in Ding (1986) and Li-Wen (1995).

The geometrical Lorenz attractor is a model abstracted from the famous equation of Lorenz (1963) for the fluid flow of the atmosphere. There are numerous works on the geometrical Lorenz attractor, from the early ones of Guckenheimer (1976, 1980) and Williams (1977, 1980) to the recent ones, such as Araújo-Pacifico (2010).

Corollary. *The following three conditions are equivalent:*

(1) *X satisfies Axiom A plus the no-cycle condition.*

(2) *X is C^1 Ω-stable.*

(3) *X is a separated star flow.*

That (1) $\Rightarrow$ (2) is the Ω-stability theorem of Pugh-Shub (1970). That (2) $\Rightarrow$ (3) is by the Kupka-Smale theorem, a lemma of Franks (1971), and the C^1 connecting lemma. That (3) $\Rightarrow$ (1) is by Gan-Wen (2006).

Though only C^1 solutions are obtained for the stability conjectures, the relevant works have formed a striking theory in differentiable dynamical systems. As Palis (2005) commented, "the C^1 case is already illuminating of the darker realm of dynamics."

We end this section with a natural question:

Why C^1?

Essentially, the reason is that some basic perturbation problems have been solved only for the C^1 topology. We name one:

The C^r closing problem. *Let $z \in M$ be a nonwandering point of a diffeomorphism $f : M \to M$. Can f be arbitrarily well C^r approximated by $g : M \to M$ such that z is a periodic point of g?*

This problem is basic to the theory of differentiable dynamical systems, in particular to the C^r stability conjecture. Indeed, to prove the C^r stability conjecture we assume f is C^r structurally stable and prove f satisfies Axiom A, in particular, prove $\Omega(f) = \overline{\mathrm{P}(f)}$. Since f is topologically conjugate to the nearby systems, it suffices to prove there is some g in a C^r neighborhood of f such that $\Omega(g) = \overline{\mathrm{P}(g)}$. This will be a direct consequence if the C^r closing problem has a positive answer. Indeed, if the C^r closing problem is verified positively, a quick argument will show that there is a C^r residual subset $\mathcal{R} \subset \mathrm{Diff}^r(M)$ such that every $g \in \mathcal{R}$ satisfies $\Omega(g) = \overline{\mathrm{P}(g)}$. See Pugh (1967b).

At first glance the answer to the C^r closing problem seems positive and easy. It is indeed positive and easy for $r = 0$. We explain why. Let $z \in M$ be a nonwandering point, and let $\delta > 0$. There are $x \in \operatorname{int} B(z, \delta)$ and $n \geq 1$ such that $y = f^n(x) \in \operatorname{int} B(z, \delta)$. We take y to be the first return of x to $\operatorname{int} B(z, \delta)$; that is, $f^i(x) \notin \operatorname{int} B(z, \delta)$ for all $1 \leq i \leq n - 1$. See Figure 6.5.

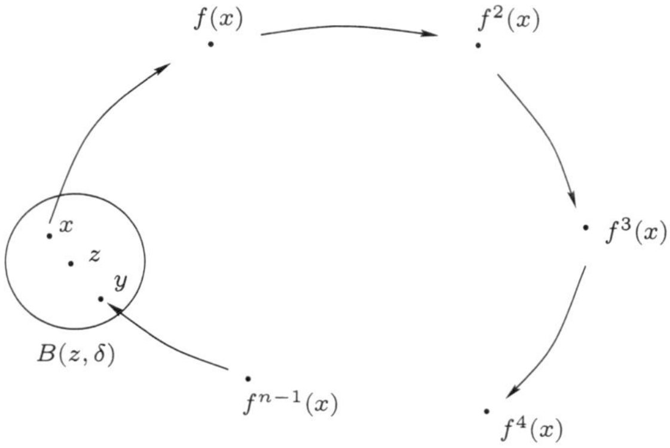

Figure 6.5. C^0 closing.

Let $h : M \to M$ be a homeomorphism such that $h = id$ outside $B(z, \delta)$ and $h(y) = x$. Let $g = h \circ f$. Then the old orbit from x to $f^{n-1}(x)$ will be unchanged; hence

$$g^n(x) = g(f^{n-1}(x)) = hf(f^{n-1}(x)) = h(y) = x.$$

That is, x becomes a periodic point of g. If δ is sufficiently small, then the C^0 distance $d^0(h, id)$ is arbitrarily small; so is $d^0(g, f)$. Another perturbation (which can be taken C^r conjugate to g) will push this periodic point x of g right onto z. Thus the C^0 closing lemma is positive and easy.

However, Peixoto observes that this argument fails for $r = 1$ (see Smale (1998) for commentary) because the C^1 distance $d^1(h, id)$ may not be small. For instance, in Figure 6.5, if x and y are near, respectively, a pair of antipodal points a and b of the sphere $\partial(B(z, \delta))$, then since $h(y) = x$ but $h(b) = b$, the first derivative of h will be of size

$$d(hy, hb)/d(y, b) = d(x, b)/d(y, b),$$

which could be arbitrarily large, no matter how small δ is.

To make $d^1(h, id)$ small, not only δ needs to be small, but also, by the mean value theorem, x and its first return y to a ball both need to be near the center of the ball, say $x \in B(y, \epsilon\delta_1)$ for arbitrarily small ϵ. See Figure 6.6. This is hardly guaranteed.

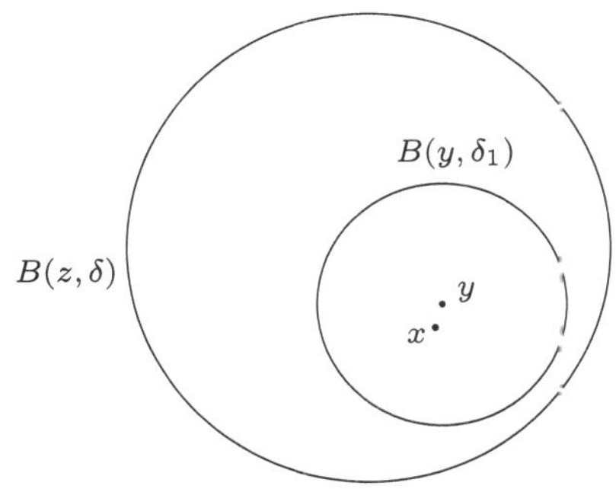

Figure 6.6. To be C^1 small, the intermediate returns between x and y must be all outside the smaller circle. This is hardly guaranteed.

With very hard constructions, Pugh (1967a) proved the following celebrated result:

Theorem (The C^1 closing lemma). *Let $z \in M$ be a nonwandering point of f. For any C^1 neighborhood $\mathcal{U}$ of f, there is $g \in \mathcal{U}$ such that z is a periodic point of g.*

Further papers on this result include Liao (1979a), Mañé (1982), Pugh-Robinson (1983), Mai (1986, 1989), Wen (1991, 1992), Arnaud (1998), Wang-Wang (1999), Rovella-Sambarino (2010).

The method used to prove the C^1 closing lemma heavily depends on a rescaling property which is strictly of the C^1 nature; that is, shrinking the C^0 size of a picture does not change the size of its first derivatives. Precisely, denote by h_δ the scalar multiplication by δ. If g is a C^1 ϵ-perturbation of id supported on the unit ball, then the rescaling $h_\delta \circ g \circ h_\delta^{-1}$ remains a C^1 ϵ-perturbation of id supported on the δ-ball, for any $0 < \delta \leq 1$. In contrast, the C^r size of this perturbation will be of order ϵ/δ^{r-1}. Thus to construct a C^1 perturbation we may leave aside the consideration of the C^0 size but focus on control of the first derivatives only. This is not the case for $r \geq 2$.

The C^r closing problem for $r \geq 2$ is open. It is one of the most important open problems in differentiable dynamical systems. At the turn of the century, Smale (1998) proposed eighteen mathematical problems for the twenty-first century, the C^r closing problem being the tenth.

Exercises

Exercise 6.1. Prove that if f satisfies Axiom A and $T_x(W^s(x)) \cap T_x(W^u(x)) = \{0\}$ for all $x \in M$, then f satisfies the no-cycle condition.

Exercise 6.2. Prove that if f satisfies Axiom A and the strong transversality condition, then f satisfies the no-cycle condition.

Exercise 6.3. Prove that a diffeomorphism that is simultaneously quasi-Anosov and satisfies the linear transversality condition must be Anosov.

Exercise 6.4. Let $f : S^1 \to S^1$ be a diffeomorphism such that $\mathrm{P}(f)$ is nonempty and consists of finitely many hyperbolic periodic points. Prove f is C^1 structurally stable.

Exercise 6.5 (C^r closing lemma)**.** Let $f : S^1 \to S^1$ be a C^r diffeomorphism with $\mathrm{P}(f) = \emptyset$, $r \geq 1$. Prove that for any C^r neighborhood $\mathcal{U}$ of f, there is $g \in \mathcal{U}$ such that $\mathrm{P}(g) \neq \emptyset$.

Exercise 6.6. Prove that, for any $r \geq 1$, if a diffeomorphism $f : S^1 \to S^1$ is C^r structurally stable, then $\mathrm{P}(f)$ is nonempty and consists of finitely many hyperbolic periodic points.

A diffeomorphism f is called *robustly expansive* if there is a C^1 neighborhood $\mathcal{U}$ of f such that every $g \in \mathcal{U}$ is expansive (see the definition of expansiveness before the statement of Theorem 4.14).

Exercise 6.7. Prove f is robustly expansive if and only if f satisfies Axiom A and $T_x(W^s(x)) \cap T_x(W^u(x)) = \{0\}$ for all $x \in M$.

Bibliography

[A] Andronov, A. and Pontryagin, L.: Systèmes grossiers, *Dokl. Akad. Nauk. SSSR* **14**, 247–251 (1937).

[An1] Anosov, D. V.: Geodesic flows on closed riemannian manifolds with negative curvature, *Proceedings of Steklov Institute of Mathematics* **90**, 1–235 (1967), Amer. Math. Soc. translation (1969).

[An2] Anosov, D. V.: On a class of invariant sets of smooth dynamical systems, *Proceedings of the Fifth International Conference on Nonlinear Oscillations*, vol. 2, 39–45. Mathematics Institute of the Ukrainian Academy of Sciences, Kiev (1970).

[Ao] Aoki, N.: The set of Axiom A diffeomorphisms with no cycles, *Bol. Soc. Brasil Mat.* **23**, 21–65 (1992).

[Ar] Araújo, V. and Pacifico, M.: *Three-dimensional Flows*, Ergebnisse der Mathematik und ihrer Grenzgebiete. 3. Folge 53. Springer, Heidelberg (2010).

[Arn1] Arnaud, M.-C.: Le "Closing lemma" en topologie C^1, Mém. Soc. Math. Fr., no. 74 (1998).

[Arn2] Arnaud, M.-C.: Création de connexions en topologie C^1, Ergod. Th. Dynam. Syst. **21**, 339–381 (2001).

[BN] Bachman, G. and Narici, L.: *Functional Analysis*, Academic Press, New York, London (1966).

[B] Banks, J., Brooks, J., Cairns, G., Davis, G., and Stacey, P.: On Devaney's definition of Chaos, *Amer. Math. Monthly* **99**, 332–334 (1992).

[BDV] Bonatti, C., Diaz, L., and Viana, M.: *Dynamics Beyond Uniform Hyperbolicity. A Global Geometric and Probabilistic Perspective*, Encyclopedia of Mathematical Sciences, 102. Mathematical Physics, III, Springer-Verlag, Berlin, Heidelberg, New York (2005).

[Bo1] Bonatti, C. and Diaz, L.: On maximal transitive sets of generic diffeomorphisms, *Publ. Math. IHES* **96**, 171–197 (2003).

[Bo2] Bonatti, C.: Generic dynamics, *Minicourse in the Summer School and Conference on Dynamical Systems at ICTP* (2004).

[Boo] Boothby, W.: *An Introduction to Differentiable Manifolds and Riemannian Geometry*, Academic Press, New York, San Francisco, London (1975).

[Bow1] Bowen, R.: *Equilibrium States and the Ergodic Theory of Anosov Diffeomorphisms*, Lecture Notes in Math., vol. 470, Springer-Verlag, New York, Heidelberg, Berlin (1975).

[Bow2] Bowen, R.: *On Axiom A Diffeomorphisms*, Regional Conference Series in Mathematics, vol. 35, Amer. Math. Soc., Providence, RI (1978).

[Br] Brin, M. and Stuck, G.: *Introduction to Dynamical Systems*, Cambridge University Press (2002).

[C] Cartwright, M. and Littlewood, J.: On nonlinear differential equations of the second order: I. The equation $\ddot{y} - k(1-y^2)\dot{y} + y = b\lambda k \cos(\lambda t + \alpha)$, k large, *J. London Math. Soc.* **20**, 180–189 (1945).

[Ch] Chillingworth, D.: *Differential Topology with a View to Applications*, Pitman Publ., London, San Francisco, Melbourne (1976).

[Ch] Churchill, R., Franke, J., and Selgrade, J.: *A geometric criterion for hyperbolicity of flows*, Proc. Amer. Math. Soc. **62**, 137–143 (1977).

[Co] Conley, C.: *Isolated Invariant Sets and Morse Index*, Regional Conference Series in Mathematics, vol. 38, Amer. Math. Soc., Providence, RI (1978).

[Cr] Crovisier, S.: Une remarque sur les ensembles hyperboliques localement maximaux, *C. R. Math. Acad. Sci. Paris* **334** (5), 401–404 (2001).

[D] Dankner, A.: On Smale's axiom A dynamical systems, *Ann. Math.* **107**, 517–553 (1978).

[De] Denjoy, A.: Sur les courbes définies par les équations différentielles a la surface du tore, *J. Math. Pure Appl.* **11**, 333–375 (1932).

[Dev] Devaney, R.: *Chaotic Dynamical Systems*, Addison-Wesley Publ. Co., New York & Reading, MA (1989).

[Di] Ding, H.: Disturbance of the homoclinic trajectory and applications, *Acta. Sci. Nat. Univ. Pekin.*, no. 1, 53–63 (1986).

[F] Fisher, T.: Hyperbolic sets that are not locally maximal, *Ergodic Theory Dynam. Systems* **26**, no. 5, 1491–1509 (2006).

[F1] Franke, J. and Selgrade, J.: Hyperbolicity and chain recurrence, *J. Diff. Eq.* **26**, 27–36 (1977).

[F2] Franke, J. and Selgrade, J.: Hyperbolicity and cycles, *Trans. Amer. Math. Soc.* **245**, 251–262 (1978).

[Fr1] Franks, J.: Necessary conditions for stability of diffeomorphisms, *Trans. Amer. Math. Soc.* **158**, 301–308 (1971).

[Fr2] Franks, J.: Absolutely structurally stable diffeomorphism, *Proc. Amer. Math. Soc.* **37**, 293–296 (1973).

[Fr3] Franks, J. and Robinson, C.: A quasi-Anosov diffeomorphism that is not Anosov, *Trans. Amer. Math. Soc.* **223**, 267–278 (1976).

[Fr4] Franks, J.: *Homology and Dynamical Systems*, Conference Board Math. Science, **49**, Amer. Math. Soc., Providence, RI (1982).

[Fr5] Franks, J.: Notes on chain recurrence and Lyapunonv functions, Personal communications (2001).

[G] Gan, S. and Wen, L.: Nonsingular star flows satisfy Axiom A and the no-cycle condition, *Invent. Math.* **164**, no. 2, 279–315 (2006).

[Gu1] Guckenheimer, J.: Absolutely Ω-stable diffeomorphisms, *Topology* **11**, 195–197 (1972).

[Gu2] Guckenheimer, J.: A strange, strange attractor, *The Hopf Bifurcation and Its Applications*, Springer-Verlag, New York, Heidelberg, Berlin (1976).

[Gu3] Guckenheimer, J. and Williams, R.: Structural stability of the Lorenz attractor, *Publ. Math. I.H.E.S.* **50**, 73–100 (1980).

[GMN] Guckenheimer, J., Moser, J., and Newhouse, N.: *Lectures on Dynamical Systems*, Progress in Math. Birkhauser **8**, 1–114 (1980).

[Gut1] Gutierrez, C.: On the C^r-closing lemma for flows on the torus T^2, *Ergod. Th. Dynam. Syst.* **6**, 45–56 (1986).

[Gut2] Gutierrez, C.: A counter-example to a C^2-closing lemma, *Ergod. Th. Dynam. Syst.* **7**, 509–530 (1987).

[Gut3] Gutierrez, C.: On C^r-closing for flows on 2-manifolds, *Nonlinearity* **13**, 1883–1888 (2000).

[Gut4] Gutierrez, C.: On the C^r-closing lemma, *Comput. Appl. Math.* **20**, 179–186 (2001).

[H1] Hartman, P.: On local homeomorphisms of Euclidean spaces, *Bol. Soc. Mat. Mexicana* **5**, 220–241 (1960).

[H2] Hartman, P.: *Ordinary Differential Equations*, Wiley, New York (1964).

[Ha1] Hayashi, S.: Diffeomorphisms in $\mathcal{F}^1$(M) satisfy Axiom A, *Ergod. Th. Dynam. Syst.* **12**, 233–253 (1992).

[Ha2] Hayashi, S.: Connecting invariant manifolds and the solution of the C^1 stability and Ω-stability conjectures for flows, *Ann. of Math.* **145**, 81–137 (1997).

[Hi1] Hirsch, M.: *Differential Topology*, Springer-Verlag, New York, Heidelberg, Berlin (1976).

[Hi3] Hirsch, M., Palis, J., Pugh, C., and Shub, M.: Neighborhoods of hyperbolic sets, *Invent. Math.* **9**, 121–134 (1970).

[Hi2] Hirsch, M. and Pugh, C.: Stable manifolds and hyperbolic sets, *Proc. Symp. Pure Math.*, Amer. Math. Soc., **14**, 133–163 (1970).

[HPS] Hirsch, M., Pugh, C., and Shub, M.: *Invariant Manifolds*, Lecture Notes in Math., vol. 583, Springer-Verlag, New York, Heidelberg, Berlin (1977).

[HS] Hirsch, M. and Smale, S.: *Differential Equations, Dynamical Systems, and Linear Algebra*, Academic Press, New York (1974).

[HY] Hocking, J. and Young, G.: *Topology*, Addision-Wesley, New York & Reading, MA (1961).

[Hu] Hu, S.: A proof of C^1 stability conjecture for 3-dimensional flows, *Trans. Amer. Math. Soc.* **342**, 753–772 (1994).

[I] Irwin, M.: *Smooth Dynamical Systems*, Academic Press (1980).

[KH] Katok, A. and Hasselblatt, B.: *Introduction to the Modern Theory of Smooth Dynamical Systems*, Cambridge University Press (1995)

[L] Levinson, N.: A second order differential equation with singular solutions, *Ann. Math.* **50**, 127–153 (1949).

[Li] Li, C. and Wen, L.: $\mathcal{X}^*$ plus Axiom A does not imply no-cycle, *J. Diff. Eq.* **119**, 395–400 (1995).

[Lia1] Liao, S.: An extension of the C^1 closing lemma, *Acta Scientiarum Naturalium Universitis Pekinensis* **3**, 1–41 (1979a).

[Lia2] Liao, S.: A basic property of a certain class of differential systems, *Acta Math. Sinica* **22**, 316–343 (1979b).

[Lia3] Liao, S.: On the stability conjecture, *Chinese Ann. of Math.* **1**, 9–30 (1980a).

[Lia4] Liao, S.: Obstruction sets I, *Acta Math. Sinica* **23**, 411–453 (1980b).

[Lia5] Liao, S.: Obstruction sets II, *Acta Scientiarum Naturalium Universitis Pekinensis* **2**, 1–36 (1981).

[Lia6] Liao, S.: Standard systems of differential equations and obstruction sets with applications to structural stability problems, *Proceedings of the 1983 Beijing Symposium on Differential Geometry and Differential Equations*, Science Press, Beijing, 65–97 (1986).

[Lia7] Liao, S.: *Qualitative Theory of Differentiable Dynamical Systems*, Science Press, China (1996).

[Lo] Lorenz, E. N.: Deterministic non-periodic flow, *J. Atmos. Sci.* **20**, 130–141 (1963).

[M1] Mai, J.: A simpler proof of C^1 closing lemma, *Scientia Sinica* **29**, 1021–1031 (1986).

[M2] Mai, J.: A simpler proof of the extended C^1 closing lemma, *Chinese Sci. Bull.* **34**, 180–184 (1989).

[Ma1] Mañé, R.: Quasi-Anosov diffeomorphisims and hyperbolic manifolds, *Trans. Amer. Math. Soc.* **229**, 351–370 (1977a).

[Ma2] Mañé, R.: Characterizations of AS diffeomorphisms, *Lecture Notes in Math.*, **597**, 389–394 (1977b).

[Ma3] Mañé, R.: An ergodic closing lemma, *Ann. of Math.* **116**, 503–540 (1982).

[Ma4] Mañé, R.: *Ergodic Theory and Differentiable Dynamics*, Springer-Verlag, New York, Heidelberg, Berlin (1987).

[Ma5] Mañé, R.: A proof of the C^1 stability conjecture, *Publ. Math. IHES* **66**, 161–210 (1988).

[Mas] Massey, W.: *Algebraic Topology: An Introduction*, Harcourt-Brace-World (1967).

[Mat] Mather, J.: Appendix to the paper of Smale, Differentiable dynamical systems, *Bull. Amer. Math. Soc.* **73**, 747–817 (1967).

[Me1] de Melo, W.: Structural stability of diffeomorphisms on two manifolds, *Invent. Math.* **21**, 233–246 (1973).

[Me2] de Melo, W. and Van Strein, S. J.: *One-Dimensional Dynamics*, Springer-Verlag, New York, Heidelberg, Berlin (1993).

[Mo] Moser, J.: On a theorem of Anosov, *J. Diff. Eq.* **5**, 411–440 (1969).

[Mu1] Munkres, J.: *Topology, Second Edition*, Prentice Hall, Englewood Cliffs, NJ (2000).

[Mu2] Munkres, J.: *Elements of Algebraic Topology*, Addison-Wesley Publishing Company, California, Massachusetts, London, Amsterdam, Ontario, Sydney (1984).

[N1] Newhouse, S.: Hyperbolic limit sets, *Trans. Amer. Math. Soc.* **167**, 125–150 (1972).

[N2] Newhouse, S.: Lectures on dynamical systems, *Dynamical Systems, Progress in Math.*, Birkhauser (1980).

[Ni] Nitecki, Z.: *Differentiable Dynamics*, M.I.T. Press, Cambridge (1971).

[P1] Palis, J.: On Morse-Smale dynamical systems, *Topology* 8 (1968), 385–404.

[P2] Palis, J. and Smale, S.: Structural stability theorems, *Proc. Symp. Pure Math.*, Amer. Math. Soc., **14**, 223–232 (1970).

[P3] Palis, J: A note on Ω-stability, *Proc. Symp. Pure Math., Amer. Math. Soc.*, **14**, 221–222 (1970).

[P4] Palis, J.: On the C^1 Ω-stability conjecture, *Publ. Math. IHES* **66**, 211–215 (1988).

[P5] Palis, J.: A global view of dynamics and a conjecture of the denseness of finitude of attractors, *Géométrie complexe et systémes dynamiques* (Orsay 1995), *Astérisque*, **261**, 335–347 (2000).

[P6] Palis, J.: A global perspective for non-conservative dynamics, *Ann. Inst. H. Poincaré, Anal. Non Linéaire* **22**, 485–507 (2005).

[PM] Palis, J. and de Melo, W.: *Geometric Theory of Dynamical Systems, An Introduction*, Springer-Verlag (1982).

[PT] Palis, J. and Takens, F.: *Hyperbolicity and Sensitive Chaotic Dynamics at Homoclinic Bifurcations*, Cambridge University Press, Cambridge, New York (1993).

[Pe] Peixoto, M.: Structural stability on two-dimensional manifolds, Topology **1**, 101–120 (1962).

[Pi] Pilyugin, S.: *Shadowing in Dynamical Systems*, Lecture Notes in Math., **1706**, Springer-Verlag (1999).

[Pl] Pliss, V. A.: A hypothesis due to Smale, Diff. Eq. **8**, 203–214 (1972).

[Po] Pollicott, M.: *Lectures on Ergodic Theory and Pesin Theory on Compact Manifolds*, London Math. Soc., Lecture Notes Series, no. 180, Cambrigde University Press, Cambridge, New York (1993).

[Pu1] Pugh, C.: The closing lemma, *Amer. J. Math.* **89**, 956–1009 (1967a).

[Pu2] Pugh, C.: An improved closing lemma and a general density theorem, *Amer. J. Math.* **89**, 1010–1021 (1967b).

[Pu3] Pugh, C.: On a theorem of Hartman, *Amer. J. Math.* **91**, 363–367 (1969).

[Pu4] Pugh, C. and Shub, M.: Ω-stability for flows, *Invent. Math.* **11**, 150–158 (1970).

[Pu5] Pugh, C. and Robinson, C.: The C^1 closing lemma, including Hamiltonians, *Ergod. Th. Dynam. Syst.* **3**, 261–313 (1983).

[R] Robbin, J.: A structural stability theorem, *Ann. Math.* **94**, 447–493 (1971).

[Ro1] Robinson, C.: C^r structural stability implies Kupka-Smale, *Dynamical Systems* (edited by M. M. Peixoto), Academic Press, New York, 443–449 (1973).

[Ro2] Robinson, C.: Structural stability of vector fields, *Ann. Math.* **99**, 154–174 (1974).

[Ro3] Robinson, C.: Structural stability of C^1 flows, *Lecture Notes in Math.*, vol. 468, Springer-Verlag, New York, Heidelberg, Berlin, 262–277 (1975).

[Ro4] Robinson, C.: Structural stability of C^1 diffeomorphisms, *J. Diff. Eq.* **22**, 28–73 (1976).

[Ro5] Robinson, C.: *Dynamical Systems, Stability, Symbolic Dynamics, and Chaos*, CRC Press (1995).

[Rov] Rovella, A. and Sambarino, M.: The C^1 closing lemma for generic C^1 endomorphisms, *Ann. Inst. H. Poincare Anal. Non Lineaire*, **27**, no. 6, 1461–1469 (2010).

[Ru] Rudin, W.: *Principles of Mathematical Analysis*, McGraw-Hill Co., Inc. (1976).

[S] Sacker, R. and Sell, G.: Existence of dichotomies and invariant splittings for linear differential systems, *J. Diff. Eq.* **15**, 429–458 (1974).

[Sa] Sannami, A.: The stability theorems for discrete dynamical systems on two-dimensional manifolds, *Nagoya Math. J.* **90** (1983) (announcement and summary: Proc. Japan Acad. Ser. A Math. Sci. **57**, 403–407 (1981)).

[Se] Selgrade, J.: Isolated invariant sets for flows on vector bundles, *Trans. Amer. Math. Soc.* **203**, 359–390 (1975).

[Sh] Shub, M.: *Global Stability of Dynamical Systems*, Springer-Verlag, New York, Berlin, Heidelberg, London, Paris, Tokyo (1987).

[Sm1] Smale, S.: Diffeomorphisms with many periodic points, *Differential and Combinatorial Topology*, Princeton Univ. Press, Princeton, NJ, 63–80 (1965).

[Sm2] Smale, S.: Strucutrally stable systems are not dense, *Amer. J. Math.* **88**, 491–496 (1966).

[Sm3] Smale, S.: Differentiable dynamical systems, *Bull. Amer. Math. Soc.* **73**, 747–817 (1967).

[Sm4] Smale, S.: The Ω-stability theorem, *Proc. Symp. Pure Math.*, Amer. Math. Soc., **14**, 289–297 (1970).

[Sm5] Smale, S.: *The Mathematics of Time: Essays on Dynamical Systems, Economic Processes, and Related Topics*, Springer-Verlag, New York, Heidelberg, Berlin (1980).

[Sm6] Smale, S.: Mathematical problems for the next century, *The Mathematical Intelligencer* **20**, no. 2, 7–15 (1998).

[Sz] Szlenk, W.: *An Introduction to the Theory of Smooth Dynamical Systems*, John Wiley & Sons (1984).

[W] Walters, P.: *An Introduction to Ergodic Theory*, Springer-Verlag, New York, Heidelberg, Berlin (1982).

[Wa] Wang, X. and Wang, D.: The C^1 closing lemma for non-singular endomorphisms equivariant under free actions of finite groups, *Trans. Amer. Math. Soc.* **351**, no. 10, 4173–4182 (1999).

[We1] Wen, L.: The C^1 closing lemma for non-singular endomorphisms, *Ergod. Th. Dynam. Syst.* **11**, 393–412 (1991).

[We2] Wen, L.: The C^1 closing lemma for endomorphisms with finitely many singularities, *Proc. Amer. Math. Soc.* **114**, 217–223 (1992).

[We3] Wen, L.: On the C^1 stability conjecture for flows, *J. Differential Equations* **129**, no. 2, 334–357 (1996).

[Wh] White, W.: An Anosov translation, *Dynamical Systems* (Proc. Sympos., Univ. Bahia, Salvador), Academic Press, New York, 667–670 (1973).

[Wi1] Williams, R.: One-dimensional nonwandering sets, *Topology* **6**, 473–487 (1967).

[Wi2] Williams, R.: Expanding attractors, *Publ. Math. I.H.E.S.* **43**, 169–203 (1974).

[Wi3] Williams, R.: The structure of Lorenz attractors, Lecture Notes in Math., **615**, 94–112 (1977).

[Wi4] Williams, R.: Structure of Lorenz attractors, *Publ. Math. I.H.E.S.* **50**, 59–72 (1980).

[Z] Zhang, Z.: *Principle of Differentiable Dynamical Systems*, China Science Press (in Chinese) (1986).

[ZQ] Zhang, J. and Qian, M.: *An Introduction to Differentiable Dynamical Systems*, Peking University Press (in Chinese) (1991).

[ZH] Zhu, D. and Han, M.: *Smooth Dynamical Systems*, East China Normal University Press (in Chinese) (1993).

Index

Selected Published Titles in This Series

173 **Lan Wen,** Differentiable Dynamical Systems, 2016
171 **Qing Han,** Nonlinear Elliptic Equations of the Second Order, 2016
170 **Donald Yau,** Colored Operads, 2016
169 **András Vasy,** Partial Differential Equations, 2015
168 **Michael Aizenman and Simone Warzel,** Random Operators, 2015
167 **John C. Neu,** Singular Perturbation in the Physical Sciences, 2015
166 **Alberto Torchinsky,** Problems in Real and Functional Analysis, 2015
165 **Joseph J. Rotman,** Advanced Modern Algebra: Third Edition, Part 1, 2015
164 **Terence Tao,** Expansion in Finite Simple Groups of Lie Type, 2015
163 **Gérald Tenenbaum,** Introduction to Analytic and Probabilistic Number Theory, Third Edition, 2015
162 **Firas Rassoul-Agha and Timo Seppäläinen,** A Course on Large Deviations with an Introduction to Gibbs Measures, 2015
161 **Diane Maclagan and Bernd Sturmfels,** Introduction to Tropical Geometry, 2015
160 **Marius Overholt,** A Course in Analytic Number Theory, 2014
159 **John R. Faulkner,** The Role of Nonassociative Algebra in Projective Geometry, 2014
158 **Fritz Colonius and Wolfgang Kliemann,** Dynamical Systems and Linear Algebra, 2014
157 **Gerald Teschl,** Mathematical Methods in Quantum Mechanics: With Applications to Schrödinger Operators, Second Edition, 2014
156 **Markus Haase,** Functional Analysis, 2014
155 **Emmanuel Kowalski,** An Introduction to the Representation Theory of Groups, 2014
154 **Wilhelm Schlag,** A Course in Complex Analysis and Riemann Surfaces, 2014
153 **Terence Tao,** Hilbert's Fifth Problem and Related Topics, 2014
152 **Gábor Székelyhidi,** An Introduction to Extremal Kähler Metrics, 2014
151 **Jennifer Schultens,** Introduction to 3-Manifolds, 2014
150 **Joe Diestel and Angela Spalsbury,** The Joys of Haar Measure, 2013
149 **Daniel W. Stroock,** Mathematics of Probability, 2013
148 **Luis Barreira and Yakov Pesin,** Introduction to Smooth Ergodic Theory, 2013
147 **Xingzhi Zhan,** Matrix Theory, 2013
146 **Aaron N. Siegel,** Combinatorial Game Theory, 2013
145 **Charles A. Weibel,** The K-book, 2013
144 **Shun-Jen Cheng and Weiqiang Wang,** Dualities and Representations of Lie Superalgebras, 2012
143 **Alberto Bressan,** Lecture Notes on Functional Analysis, 2013
142 **Terence Tao,** Higher Order Fourier Analysis, 2012
141 **John B. Conway,** A Course in Abstract Analysis, 2012
140 **Gerald Teschl,** Ordinary Differential Equations and Dynamical Systems, 2012
139 **John B. Walsh,** Knowing the Odds, 2012
138 **Maciej Zworski,** Semiclassical Analysis, 2012
137 **Luis Barreira and Claudia Valls,** Ordinary Differential Equations, 2012
136 **Arshak Petrosyan, Henrik Shahgholian, and Nina Uraltseva,** Regularity of Free Boundaries in Obstacle-Type Problems, 2012
135 **Pascal Cherrier and Albert Milani,** Linear and Quasi-linear Evolution Equations in Hilbert Spaces, 2012
134 **Jean-Marie De Koninck and Florian Luca,** Analytic Number Theory, 2012

For a complete list of titles in this series, visit the
AMS Bookstore at **www.ams.org/bookstore/gsmseries/**.